Planar Slow Wave Structure
Traveling Wave Tubes

Design, fabrication and experiment

Online at: https://doi.org/10.1088/978-0-7503-5452-3

Series in Electromagnetics and Metamaterials

Series Editor
Akhlesh Lakhtakia, *Pennsylvania State University*

Series in Electromagnetics and Metamaterials

The Series on Electromagnetics and Metamaterials, published by IOP, is an innovative and authoritative source of information on a fundamental science that has been enabling a multitude of transformative technologies for two centuries and more. The electromagnetic spectrum extends from millihertz waves to microwaves to terahertz radiation to ultraviolet light and even soft x-rays. In each spectral regime, different classes of materials have different kinds of electromagnetic response characteristics. Much has been discovered and much has been technologically exploited, but even more remains to be discovered and even more remains to be put to use for diverse applications.

Electromagnetics is an evermore vibrant arena of techno-scientific research. This is amply exemplified by the huge current interest in metamaterials. By virtue of carefully designed and engineered morphology, metamaterials exhibit response characteristics that are either completely absent or muted in their constituent materials.

Each book in the series offers an extended essay on a foundational topic: an emerging topic; a currently hot topic; and/or a tool for metrology, design, and application. Ranging from 60 to 120 pages, books are written by internationally renowned experts who have been charged with making the content not only authoritative, but also easy to understand, thereby offering more synthesis and depth than a typical review article in a journal.

Illustrated in full color for both ebook and printed copies, these short books are easily searchable in the ebook format. The series is thus more modular and dynamic than traditional handbooks and more coherent than contributed volumes.

This series is edited by Akhlesh Lakhtakia, the Charles Godfrey Binder (Endowed) Professor of Engineering Science and Mechanics at the Pennsylvania State University. Initial topics targeted in the series include symmetries of Maxwell equations, homogenization of bianisotropic materials, metamaterials and metasurfaces, transformation optics, nanophotonics for medicine and biology, single photons, radiation sources, optical bolometry, magnetic resonance imaging, and detection and imaging of buried objects. Additional topic suggestions are welcomed and will be promptly considered and decided upon. As part of the IOP digital library, students and professors at purchased institutions will have unlimited access to the ebooks for classroom and research usage.

A full list of the titles published in this series can be found at: https://iopscience.iop.org/bookListInfo/iop-series-on-electromagnetics-and-metamaterials#series.

Planar Slow Wave Structure Traveling Wave Tubes

Design, fabrication and experiment

Yubin Gong
Shaomeng Wang
National Key Laboratory of Science and Technology on Vacuum Electronics,
School of Electronic Science and Engineering, University of Electronic Science
and Technology of China (UESTC), Chengdu, China

IOP Publishing, Bristol, UK

ISBN   978-0-7503-5452-3 (ebook)
ISBN   978-0-7503-5450-9 (print)
ISBN   978-0-7503-5453-0 (myPrint)
ISBN   978-0-7503-5451-6 (mobi)

DOI   10.1088/978-0-7503-5452-3

Version: 20241001

IOP ebooks

British Library Cataloguing-in-Publication Data: A catalogue record for this book is available from the British Library.

Published by IOP Publishing, wholly owned by The Institute of Physics, London

IOP Publishing, No.2 The Distillery, Glassfields, Avon Street, Bristol, BS2 0GR, UK

US Office: IOP Publishing, Inc., 190 North Independence Mall West, Suite 601, Philadelphia, PA 19106, USA

Dedicated to colleagues and students who have been, are or will be engaged in traveling wave tube amplifier developments and applications.

Contents

Preface

The traveling wave tube (TWT) is a member of the vacuum electronic device (VED) family, which provides an interaction site for and realizes energy exchange between a high energy direct current (DC) electron beam and a high frequency electromagnetic wave. Specifically, a TWT employs a slow wave structure (SWS) as the beam–wave interaction site, on which the electromagnetic wave can propagate with a phase velocity smaller than the speed of light in free space. By selecting the form and structure parameters of the transmission line or waveguide used to construct the SWS, a broad operating bandwidth can be achieved. Thus, the TWT has been widely applied in communications, broadcast, detection, etc. There have also been books talking about the TWT, such as the classic *Principles of Traveling Wave Tubes* by A S Gilmour.

Since the invention of the TWT, helix has been the dominant SWS due to its weak dispersion characteristics and high beam–wave interaction impedance. It is not difficult to achieve a 2~3 octave operating bandwidth for a helix TWT at the centimeter wavelength bands. However, things have changed in recent years. As the spectrum resources being fully developed at the low frequency wave band, development of electromagnetic wave amplifiers operating at millimeter wave and higher bands is becoming a foremost problem for further applications. The planar SWS TWTs, which cooperate with sheet electron beams, have the ability to settle the power limitation caused by the size reduction in traditional cylindrical-beam TWTs. In particular, being competitive candidate millimeter wave and terahertz wave amplifiers for 5G/6G communications, the planar SWS TWTs are attracting more and more attention. Yet there is no book specifically introducing the key points in the development of such a device.

We aim to provide solutions for the following questions in this book: the basic concepts of planar TWT; SWS TWTs should be investigated; how a planar SWS should be designed; how the electro-optics system (EOS) for a planar SWS should be designed; the fabrication of planar SWS using advanced techniques; assembly of the planar SWS TWT; test of the components and entire planar SWS TWT. TWT researchers or students may learn how to design or improve a wideband, high efficiency, miniature millimeter wave and terahertz wave amplifiers. System engineering can learn about the ability and operating guide of planar SWS TWTs, which will help them to develop better systems.

Chapter 1 talks about the concept and proposal of the planar TWT, and the target problems to solve. Chapters 2 and 3 introduce the topologies, high frequency characteristics, and advantages of different kinds planar SWS and EOS, respectively. In chapter 4, some advanced manufacture techniques which could be used in the development of planar TWTs are discussed. Chapter 5 demonstrates the fixtures and assembly of different kinds of planar TWTs, and chapter 6 demonstrates some system tests results. The last chapter is about the perspective and discussion.

Acknoweldgements

We wish to thank my previous and current group members whose contributions to the works presented have made this book possible, they are Dr Duo Xu, Dr Hexin Wang, Dr Xinyi Li, Dr Tenglong He, Mr Jingyu Guo, Mr Yang Dong, Miss Yuxin Wang, Mr Youfeng Yang, Mr Yuxin Wang, Miss Mi Tian. Ms Qingying Yi helped to check the full text. We would also like to thank the IOP Publishing eBooks Production Team, especially the Coordinator Betty Barber, who has worked diligently to bring this book to fruition.

Author biographies

Yubin Gong

Professor Yubin Gong Received his PhD degree in physical electronics from University of Electronic Science and Technology of China (UESTC) in 1998. Now he is a distinguished full professor in UESTC. He is a member of the IEEE International Technical Committee on Vacuum Electronics and the general president of 24th International Vacuum Electronics Conference (IVEC). His research interests focus on millimeter and terahertz wave vacuum electron devices. He has published several hundred papers and holds more than 100 patents.

Shaomeng Wang

Professor Shaomeng Wang obtained his PhD in 2013 from the University of Electronic Science and Technology of China. He has been a research fellow with Nanyang Technological University, Singapore. His research interests include millimeter/terahertz wave vacuum electron devices and applications. He is a senior member of the IEEE. He has served as chair of the local organized committee, chair of technical program committee, and session chair for several international conferences. He has published more than 100 papers and holds 10 patents.

Planar Slow Wave Structure Traveling Wave Tubes
Design, fabrication and experiment
Yubin Gong and Shaomeng Wang

Chapter 1

Introduction

With the rapid development of science and technology, electromagnetic waves (EMWs) are increasingly finding application across various fields and mediums. The EMW spectrum, characterized by frequency or wavelength, has been segmented into distinct bands in accordance with radar frequency bands established by IEEE standards. Due to EMWs' short wavelength, high frequency, and robust penetration capabilities, they have found broad utilization in military applications, scientific research, and various other fields. Noteworthy examples in daily life include Bluetooth, which operates at a frequency of 2.4 GHz; wireless LAN, which functions at 2.5 GHz or 5 GHz; as well as applications in medical diagnosis and treatment, such as x-rays and gamma rays. Furthermore, EMWs play a crucial role in diverse radar systems, electronic countermeasures, and high-power microwave systems.

As the basis of all these applications, EMW power amplifiers have consistently remained the focal point of research and attention. Currently, microwave power amplifiers are predominantly categorized into two types: solid-state amplifiers and vacuum electron amplifiers. A solid-state power amplifier (SSPA) is an electronic device that achieves signal amplification by harnessing the movement of internal electrons in solids. Its rapid development has been closely tied to advancements in third-generation semiconductor technology, which offers advantages such as compact size and cost-effectiveness.

Common types of SSPA include silicon laterally diffused metal-oxide semiconductors (Si-LDMOSs), silicon bipolar junction transistors (Si-BJTs), gallium arsenide (GaAs) devices, and GaN high-electron-mobility transistors (HEMTs). Furthermore, SSPAs can reach high power through integration. For example, an SSPA made by SOPHIA in the United States can output 125 W of power at 12.5–18 GHz. Xidian University has achieved an output of 138 W in the K_u band, and power in the kilowatt range can be obtained through integration. One of the prominent advantages of SSPAs is their exceptionally low operating voltage, requiring only tens of volts to function optimally. For instance, the operating voltage of typical GaAs devices is less than 15 V. Even though the operating voltages of Si-BJTs and Si-LDMOSs are

doi:10.1088/978-0-7503-5452-3ch1

relatively higher, they still only range from 28 V to 48 V. Therefore, they have found extensive application in civil communications, low-frequency transmitters, and various other domains. A clear trend for replacing vacuum electron amplifiers is increasingly prevalent. Nevertheless, challenges persist in addressing issues such as low frequency, limited bandwidth, suboptimal efficiency, and the relatively short lifespan associated with solid-state amplifiers. Efforts to resolve these challenges are ongoing.

Vacuum electron amplifiers, also known as vacuum electron devices (VEDs), are electronic systems that use charged particles (usually electrons) to interact with EMWs in vacuum and convert the direct current (DC) energy of electrons into microwave energy to achieve power amplification. VEDs feature high power and high reliability. Due to their extensive developmental history, VED types exhibit a broad range of variations. The primary categories include: electrostatic control microwave triodes, tetrodes, dynamically controlled traveling-wave tubes (TWTs), klystrons, magnetrons, backward wave oscillators (BWOs), negative-mass-effect gyrotrons, relativistic Cherenkov devices, and free-electron lasers (FELs).

Despite facing significant challenges in recent years, SSPAs continue to possess notable advantages in terms of high bandwidth, high efficiency, reliability, and longevity. As a result, they are still extensively utilized across various industries, including the medical sector.

As a crucial member of the VED family, the TWT is the most important microwave tube in modern military equipment due to its irreplaceable high bandwidth, high power, and high efficiency. Figure 1.1 shows a comparison between the average single-tube powers of VEDs and those of solid-state devices, where the EIK represents extended interaction klystron, BWO represents backward wave oscillator, FEL means free electron laser, IMPATT means impact avalanche transit time diode, MMIC means millimeter wave integrated circuit, RTD means resonant

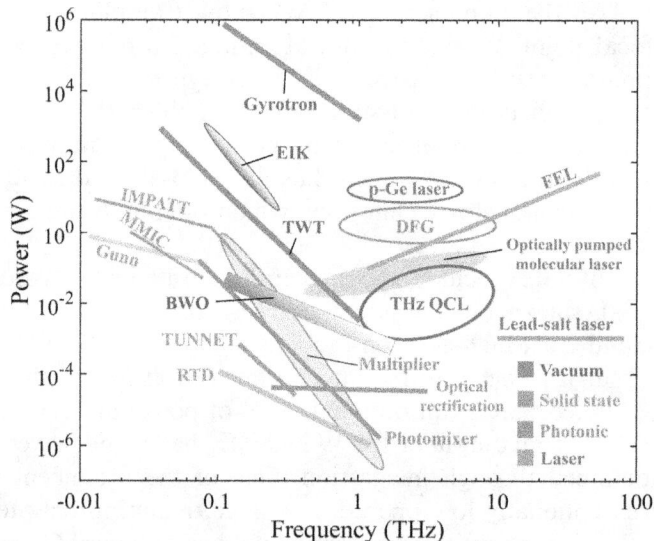

Figure 1.1. Average single-tube powers of vacuum electron devices and solid-state devices.

tunneling diode. It is evident that VEDs continue to maintain substantial advantages in terms of both working frequency and output power.

1.1 Conventional traveling-wave tubes

A TWT comprises two primary components: an electron beam and a high-frequency EMW; in this respect, it is similar to the majority of VEDs. Therefore, a TWT can be divided into two major subsystems: an electro-optical system (EOS) used to generate, form, bunch, and collect an electron beam, and a slow-wave system used to feed, transmit, and output a high-frequency electromagnetic field. The typical structure of a TWT is shown in figure 1.2.

1.1.1 Electro-optical systems

As a means of converting DC power energy into high-frequency electromagnetic energy, a strong electron beam is indispensable in a VED. The shape and internal current distribution of the beam fundamentally determine the efficiency, gain, stability, and noise characteristics of the TWT. Therefore, the system that generates, forms, and focuses the electron beam greatly affects the development of new devices. The system that performs these functions is the EOS.

A standard EOS comprises several key components. First, an electron gun is employed to emit electrons, facilitating the formation of the electron beam and accelerating it to the desired speed. Second, a focusing system is integrated to maintain the electron beam's shape, preventing interception by a slow-wave structure (SWS). Finally, a collector is implemented to gather the electrons that have completed their task, recycling a portion of the DC energy to enhance the overall efficiency of the tube.

In today's widely used TWTs, the electron beam currents, sizes, and shapes vary. The currents range from a fraction of a microampere to hundreds of milliamps, and the voltages range from thousands to tens of thousands of volts. The shape of the

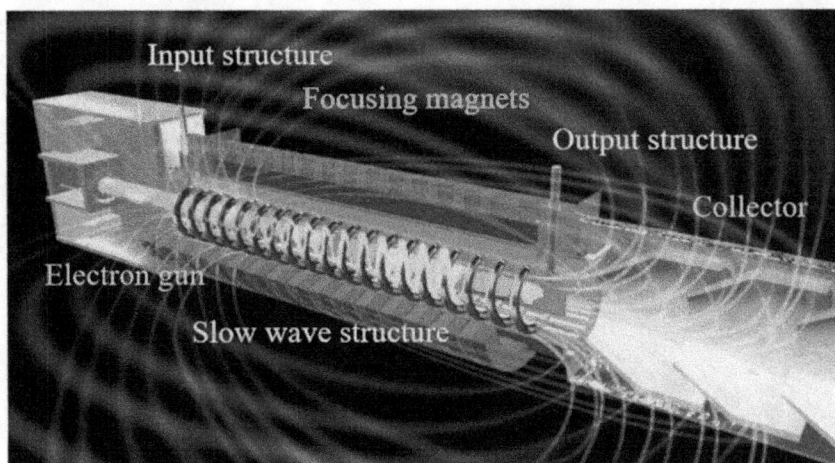

Figure 1.2. Structural diagram of a helical TWT.

electron beam is usually designed to match the size and shape of the beam tunnel of the SWS. The shape of the electron beam can be categorized into two types: cylindrical electron beams (including hollow circular electron beams) and sheet electron beams.

1.1.1.1 Cylindrical electro-optical systems
A cylindrical electron beam is an electron beam with a circular (or annular) cross-section. It possesses an axisymmetric structure, offering the advantages of simplicity of design, ease of focusing, and convenient processing. It has been used in both helical TWTs and coupled-cavity TWTs, which are widely used in radar systems and satellite communications.

1.1.1.1.1 Circular electron guns
Circular electron guns are categorized into parallel electron beam guns and spherical convergent electron beam guns, as shown in figure 1.3. These beams can be obtained using a plate diode and a spherical diode, respectively. While the calculation and structure of the flat electron beam gun are straightforward, it is important to note its evident disadvantages. These drawbacks include the necessity for a stronger magnetic field to converge the beam, limitations on the beam current density imposed by the cathode emission capacity, and the shorter lifespan of the cathode. Consequently, the utilization of spherical convergent electron beam guns is more prevalent due to their practical advantages.

The cathode serves as the source of electron emission. The main types of cathodes include thermal emission cathodes, field-induced emission cathodes, and light-induced emission cathodes. Most TWTs use a thermal emission cathode. The emission current density and cathode lifespan are crucial parameters used to assess cathode quality. In various applications, there is a significant demand for cathodes to exhibit both a high current density and an extended lifespan. Maximizing these attributes is essential to enhancing the reliability of devices while simultaneously reducing costs.

The advancement of modern surface analysis technology has significantly enhanced cathode performance. Using the oxide cathode as an illustration, its development since its inception in 1904 has reached a remarkable level. The DC emission capacity of oxide cathodes has risen from several mA cm^{-2} to tens of A cm^{-2}. In recent

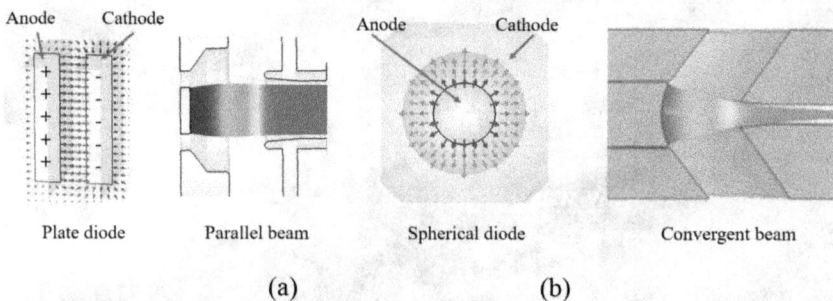

Figure 1.3. Schematic diagrams of two round electron guns. (a) A gun based on a plate diode—a parallel electron beam gun and (b) a gun based on a spherical diode—a convergent electron beam gun.

years, researchers have shown extensive interest in scandate cathodes due to their low escape work function and robust emission capabilities. The emission current density of scandate cathodes has reached 400 A cm^{-2}. In practical applications, such as millimeter-wave TWTs, scandate cathodes can maintain operational integrity for thousands of hours, even under conditions of limited current density. Notably, the other highest reported emission current density is associated with cathodes made of $Ba_xSr_{1-x}HfO_3$ salt, which reached 1014 A cm^{-2}.

The cold cathode is an important development trend in the field of high-current-density cathodes. Currently, the main types of field-induced emission cathodes include the Spindt cathode and the carbon nanotube (CNT) cathode. The highest emission current density of the former can reach 2000 A cm^{-2}, and the latter can reach 12 A cm^{-2} under limited current conditions.

The focusing electrode is an electrode used to form a parallel or converging electron beam. It usually has the same potential as the cathode and is placed at an angle of 67.5° relative to the cathode, which is called the Pierce angle. The main role of the anode is to pass DC energy to electrons. In practical applications, the shapes of the beam-forming electrode and the anode are tailored differently to fulfill distinct electron beam requirements. In addition, for enhanced control over the electron beam, it may be necessary to introduce a grid or a second anode. Figure 1.4 shows the profile of an electron gun controlled by a focusing pole and an anode.

1.1.1.1.2 Circular electron beam-focusing systems

As the electron beams used in VEDs have a strong internal space charge force and diverge rapidly during transmission, an external focusing force is needed to constrain (focus) the electron beam. Common means of focusing include magnetic focusing and electrostatic focusing. Magnetic focusing relies on the Lorentz force, wherein the direction of movement interacts with either a congruent or opposing magnetic field to generate a focusing force. This method can be categorized into uniform magnetic field focusing and periodic magnetic field focusing, depending on the variation in the magnetic field. The establishment of a focusing field can be achieved through the use of electrified spiral coil packages or permanent magnets.

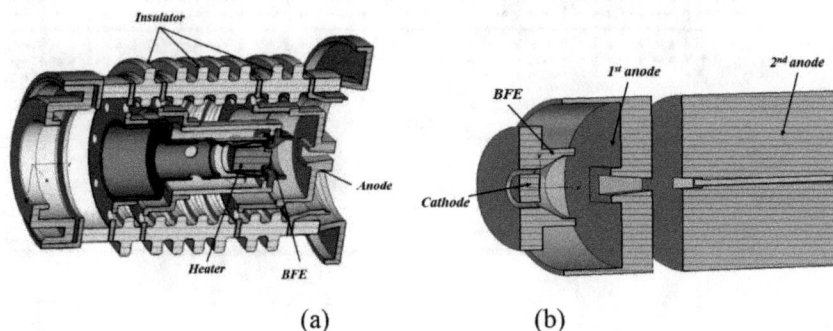

(a)   (b)

Figure 1.4. Two electron beam control structures. (a) Beam-forming electrode (BFE) controlled electron gun; (b) anode-controlled electron gun.

The electrified helical coil package offers several advantages, including ease of processing, convenient adjustment of the magnetic field value, and the theoretical capability to attain any desired shape and value of the magnetic field. In practical applications, piecewise wound energized wire packages are often used to generate magnetic fields with a given degree of uniformity, as shown in figure 1.5. However, a drawback associated with the helical wire package is its heightened demand for DC energy consumption, coupled with an increase in both volume and weight, particularly when a larger magnetic field is necessitated. Therefore, in high-power vacuum amplifiers, periodic permanent magnets are frequently employed to generate a focusing magnetic field, so as to reduce the size and weight of the device.

A periodic permanent magnet is a focusing system that links a series of magnets, typically ranging from about 20 to 40, in series. This arrangement is designed to generate an alternating periodic magnetic field of the desired length (figure 1.6).

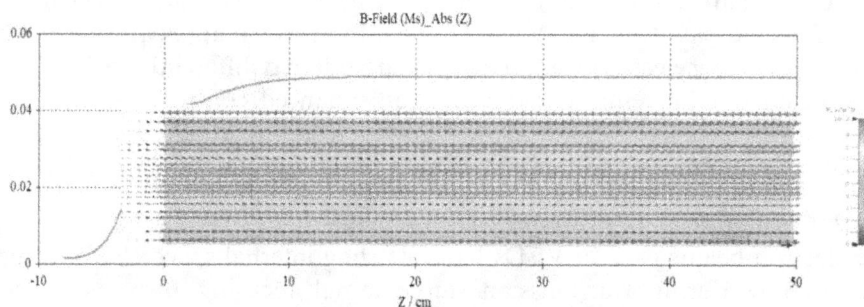

Figure 1.5. Magnetic field generated by three segments of a spiral wire pack.

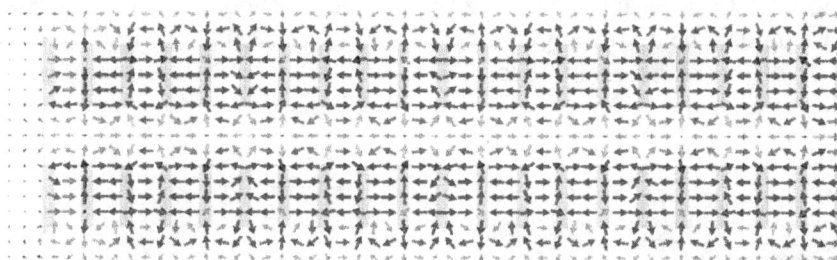

(a)

(b)

Figure 1.6. (a) Magnetic field generated by periodic permanent magnets; (b) electron-beam envelope.

In contrast to the electrified helical wire coil package, permanent magnets offer distinct advantages, such as zero energy consumption and heightened reliability. Furthermore, the utilization of permanent magnets has the potential to significantly reduce the weight of the focusing system by one to three orders of magnitude. Permanent magnets are well-suited for medium- and low-power TWTs that have long electron beams. Today, the pertinent theory, design, and application aspects have reached a relatively advanced stage, and a wealth of literature is available.

Electrostatic focusing stands out as a practical method for beam focusing. It allows bulky spiral packages to be abandoned, typically resulting in savings in auxiliary power and DC power consumption. This, in turn, contributes to an overall enhancement in the tube's efficiency. Compared with the periodic permanent magnet focusing mode, TWTs that use the electrostatic focusing mode are not affected by the ambient temperature. In addition, the electrostatic focusing method can greatly prolong the life of the cathode and restrain parasitic ion oscillation. To date, TWTs, BWOs, and klystrons using electrostatic focusing have been developed successfully. There are three basic methods of electrostatic focusing: centrifugal–electrostatic, periodic electrostatic, and individual strong electron lens focusing.

Periodic electrostatic focusing involves the utilization of a periodic electrostatic field to concentrate the electron beam. This method commonly includes the application of a double-helix field (figure 1.7) and periodic electrostatic focusing performed by specially shaped electrodes.

Periodic electrostatic focusing with specially shaped electrodes is a focusing technique that initially establishes a periodic distribution of potential along the electron beam. Subsequently, corresponding electrodes are designed to eliminate transverse electric field component. The methods used to design these electrodes include calculation methods and electrolytic cell methods.

1.1.1.1.3 *Circular electron beam collectors*

In a vacuum device, the collector is responsible for absorbing all the energy carried by the electrons that have done the work, thereby contributing to an overall

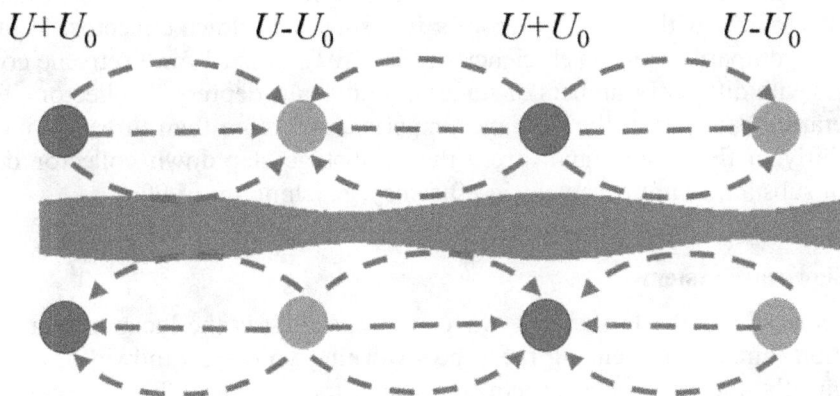

Figure 1.7. Schematic diagram of electron motion in the electric field of a double helix.

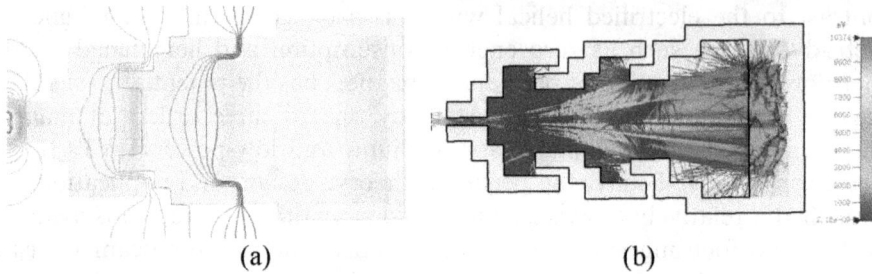

(a)   (b)

Figure 1.8. (a) Equi-static surfaces and (b) electron trajectories in a multistage depressed collector.

efficiency improvement of the device. The focal point of the collector design lies in establishing the precise trajectory of electrons within the collector and selecting an appropriate surface shape for electron bombardment. The step-down collector is a frequently employed design in modern TWTs. It operates at a voltage lower than the potential of the tube body, effectively decreasing the speed of electrons as they enter the collector. This reduction in speed aims to minimize the heat generated by the conversion of electronic kinetic energy, facilitate the recovery of electronic energy, and consequently enhance the overall efficiency of the tube. However, in practical applications, considering the substantial generation of secondary electrons on the collector surface due to electron bombardment, it is essential to limit the potential drop of the collector. An excessively high potential drop could result in a significant return current between the slow-wave system and the collector, thereby impeding the efficiency enhancement of the depressed-collector tube. A strategy for further enhancing efficiency involves the implementation of a multistage step-down collector. This method utilizes multiple electrodes with varying potentials to decelerate electrons. It allows for the separate collection of electrons and secondary electrons with different energies, thereby contributing to improved overall efficiency.

Figure 1.8 shows the isophase surfaces and electronic trajectories of a four-stage step-down collector. Its compact size makes it suitable for airborne systems with stringent size and weight requirements.

The results indicate a notable enhancement in device efficiency, which increases from 24% to 55% with the adoption of a two-stage step-down collector. Figure 1.9 presents a comparison of the efficiency of the TWTs utilized for electronic counter-measures with different numbers of stages in multistage depressed collectors. Within the operating bandwidth, the efficiency experiences a more than threefold increase. Particularly in the small signal area, the multistage step-down collector demonstrates a substantial improvement in efficiency, reaching up to 90%.

1.1.2 Slow-wave systems

As the core of the TWT, the slow-wave system serves as the locus of beam–wave interaction, directly influencing the tube's working voltage, bandwidth, and gain. Consequently, the slow-wave system has been a focal point of TWT research since its inception. To date, hundreds of different SWSs have been proposed.

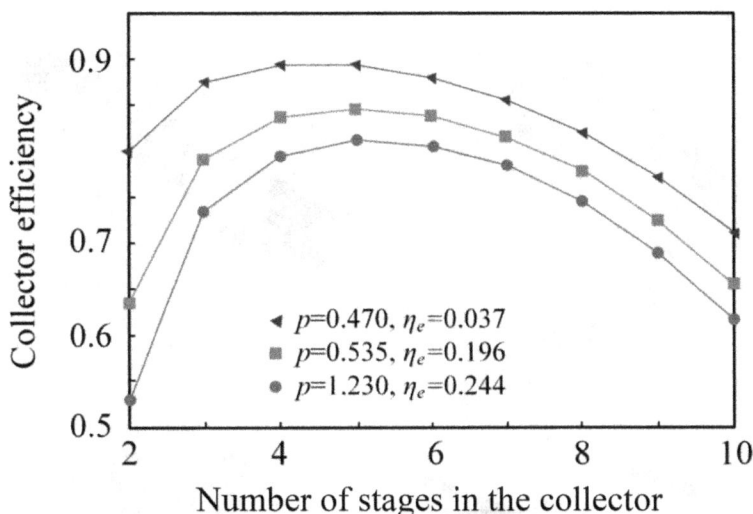

Figure 1.9. Comparison between the efficiencies of TWTs with varying numbers of stages in their depressed collectors [1]. p is perveance and η_e is electron efficiency.

1.1.2.1 Circular electron beam slow-wave systems

The circular electron beam appeared at the inception of the TWT. Therefore, the types of slow-wave systems that use circular electron beams are also the most numerous, such as the classical helix, coupled cavities, folded waveguides, and their deformed structures.

The helix stands out as the slow-wave system with the broadest bandwidth due to its notably weak dispersion characteristics. However, the practical implementation of a helical system necessitates the use of insulating materials, which can result in an increase in dispersion strength. In addition, the coupling between the current of the helix and the current of the shell also affects the coupling impedance. Therefore, it is usually necessary to load the fins to enhance the performance of helical slow-wave systems. Experimental results demonstrate that the application of T-shaped metal fin loading technology can significantly augment the small signal gain of the helical TWT, expanding it to more than two octaves with an achievable gain of 60 dB. To date, in the K_a band, the output power of a spiral TWT has reached 500 W at an efficiency of 17%.

Figure 1.10 shows some deformed helical structures, including SWSs such as the reverse double helix, the ring–rod SWS, the spiral groove waveguide, and the spiral circular waveguide with angular periodic loading. In comparison to the helix, while these structures may exhibit narrower bandwidths, they excel at suppressing backward wave oscillation and enhancing output power performance. For example, a TWT using the ring–rod slow-wave system can output 10 kW of power at a bandwidth of 8.4–10.2 GHz.

The coupled cavity is a periodically loaded slow-wave system extensively employed in high-power TWTs. Multiple cavities, such as klystrons, are typically utilized as the slow-wave system. While the bandwidths of these systems may not be

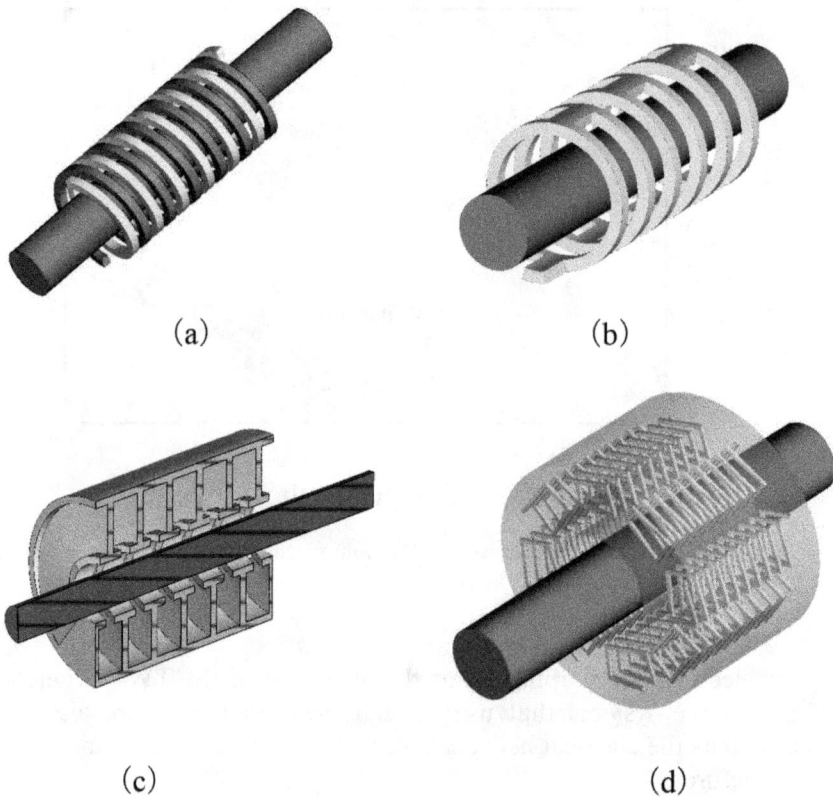

Figure 1.10. Deformed structures of spiral slow-wave systems. (a) Reverse helix [2]; (b) ring–rod slow-wave line [3]; (c) spiral waveguide [4]; and (d) helical circular waveguide with angular loading [5].

as high as those of helical systems, they excel at operating at very high voltages, enabling the attainment of significantly higher peak power compared to that of helical TWTs. An all-metal structure can indeed contribute to higher average power. Research indicates that a TWT featuring a staggered coupled slot structure can achieve an output power of 857 W in the 5 GHz region of the K_a band.

Figure 1.11 illustrates a selection of coupled-cavity slow-wave systems that exhibit commendable performance, including the Hughes structures, including the staggered coupling cavity and the double-hole coupled cavity, and the cloverleaf structure.

1.2 Problems encountered

In general, applications necessitate TWTs to be compact and potent, to operate at high frequencies, to possess a wide bandwidth, and to demonstrate high efficiency. Nevertheless, it is important to note that these performance indicators often conflict with each other in the context of TWTs. On the one hand, the adoption of high frequencies results in a diminutive beam tunnel size, necessitating the utilization of a smaller beam. Conversely, a smaller beam implies reduced DC energy under

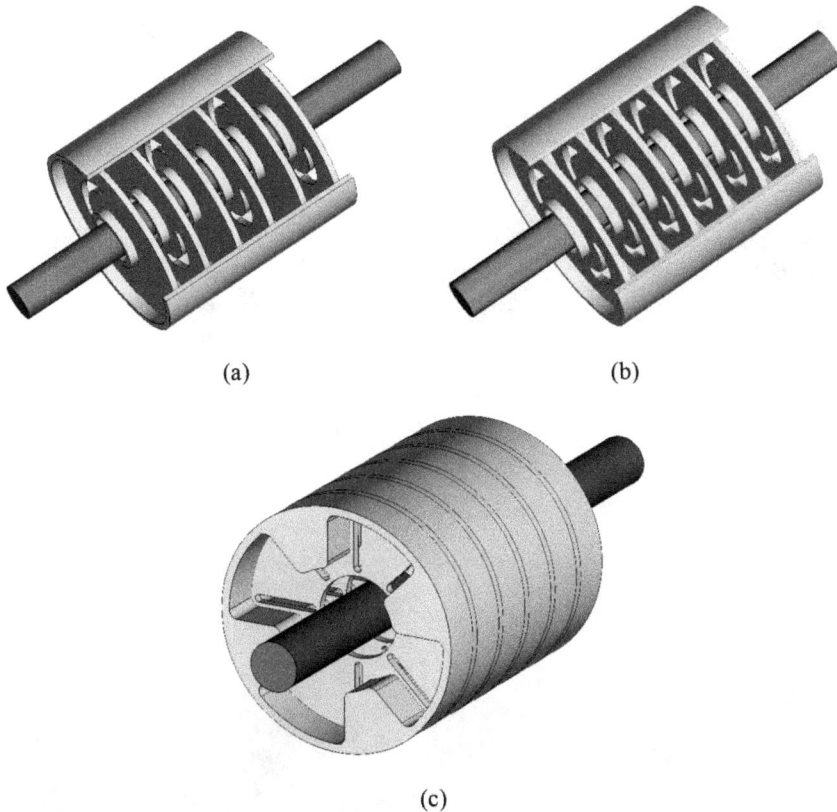

Figure 1.11. Coupled-cavity slow-wave systems. (a) Staggered coupling cavity [6]; (b) double-hole coupled cavity; and (c) cloverleaf coupling cavity [7].

identical beam voltage and current density conditions. To alleviate the constrained correlation between frequency and output power, various solutions have been proposed. On the beam side, the application of multiple, sheet, or ring electron beams can yield a greater current, thereby increasing the DC energy. On the EMW side, opting for planar SWSs or exploiting higher-order EMW modes of the SWSs yields a larger bandwidth within higher frequency bands.

References

[1] Li F, Liu X, Liu P K, Yang G J, Yi H X and Wang X S 2012 Study on estimating efficiency of multistage depressed collector in traveling wave tubes *Acta Phys. Sin.* **61** 102901

[2] Cain W N and Grow R W 1990 The effects of dielectric and metal loading on the dispersion characteristics for contrawound helix circuits used in high-power traveling-wave tubes *IEEE Trans. Electron Devices* **37** 1566–78

[3] Datta S K, Naidu V B, Rao P R R, Kumar L and Basu B 2009 Equivalent Circuit Analysis of a Ring–Bar Slow-Wave Structure for High-Power Traveling-Wave Tubes *IEEE Trans. Electron Devices* **56** 3184–90

[4] Wang W, Gong Y, Wei Y *et al* 2002 The advance of new slow-wave structure for high-power TWT *Vac. Electron.* **6** 13–8

[5] Liu Y, Gong Y, Xu J *et al* 2011 A circular waveguide slow wave structure with angular loaded helices *Chinese Patent* 201020666835 6.22.2011

[6] He F, Luo J, Zhu M and Guo W 2013 Theory, simulations, and experiments of the dispersion and interaction impedance for the double-slot coupled-cavity slow wave structure in TWT *IEEE Trans. Electron Devices* **60** 3576–83

[7] Chodorow M and Craig R A 1957 Some new circuits for high-power traveling-wave-tubes *Proc. IRE* **45** 1106–18

Chapter 2

Planar slow-wave structures

A planar slow-wave structure (SWS) is typically a structure that is uniform in at least one transverse direction, making it well-suited for 2D manufacturing technologies and sheet electron beams. In contrast to 3D SWSs, such as helical or coupled cavity designs, the planar SWS proves to be more competitive, particularly when operating in the millimeter-wave and higher frequency bands. This planar configuration can be constructed from various types of waveguides or transmission lines by periodically or quasi-periodically bending them in the axial direction.

2.1 Periodic planar slow-wave structures

Figure 2.1 displays SWSs designed to operate with sheet beams, such as the rectangular grating, the staggered double vane, the folded groove waveguide, the sinusoidal waveguide, the microstrip meander line, and the planar helix with straight edge connections.

These SWSs are capable of operating across a range of frequency bands, spanning from the V band and the W band to the G band and beyond. The computer simulation results presented in the literature demonstrate that the output power of the V-shaped folded groove waveguide surpasses 1 kW, with a corresponding gain exceeding 33 dB within the frequency range from 58 GHz to 64 GHz [1]. The staggered dual vane exhibits a peak power of 1100 W within the 86–110 GHz band and surpasses 300 W in the 132–152 GHz band [2]. In the case of sinusoidal waveguides, the peak power output exceeds 150 W within the frequency range of 200–240 GHz [3].

2.1.1 Radial beam traveling-wave tubes

Conventional traveling-wave tubes (TWTs) that use cylindrical or ring electron beams are classified as longitudinal structures; they are characterized by the following attributes: the electron beam is emitted from the cathode situated at one end of the TWT and travels in a singular direction within the Cartesian coordinate

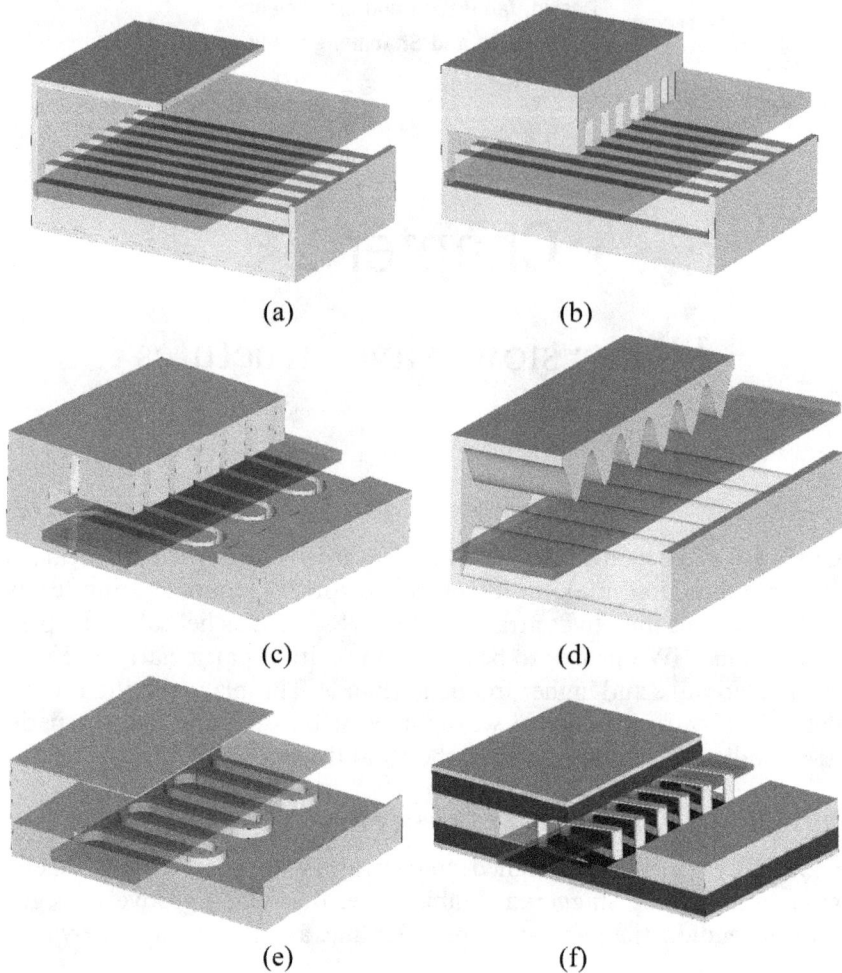

Figure 2.1. Several novel periodic SWSs. (a) The rectangular grating; (b) the staggered double vane; (c) the folded groove waveguide; (d) the sinusoidal waveguide; (e) the microstrip meander line; and (f) the planar helix with straight connections.

system toward the collector pole positioned at the opposite end. At any given moment, the frontal aspect of the electron beam forms a plane. The electromagnetic wave (EMW) travels along the lateral path of the SWS, and its longitudinal component interacts with the synchronized electron beam. To ensure efficient interaction and a workable bandwidth, the transverse dimension of a conventional slow-wave system must not be excessively large. The ratio of the transverse length to the longitudinal length in a single-period SWS directly influences the phase velocity of the EMW. This ratio, in turn, limits the possible reduction in the operating voltage of conventional TWTs. The operational requirement for voltages in the tens of thousands of volts not only poses a significant challenge in the assembly of TWTs but also imposes substantial limitations on their practical applications. In contrast, radial beam TWTs (RB-TWTs) offer the advantage of reducing the operating voltage to the range of thousands or even hundreds of volts.

An RB-TWT utilizes a radial sheet electron beam (RSEB) distinguished by the following characteristics: electrons are emitted from the side of a cylindrical (or fan-shaped) cathode positioned at the center of the TWT. The resulting RSEB travels in the radial direction, and at any given moment, the frontal aspect of the electron beam forms a circle (or an arc) centered at the cathode. The EMW travels along the angular path of the slow-wave system, and its radial component interacts and exchanges energy with the synchronous electron beam. Due to the potential for a substantial ratio between angular and radial sizes, the radial slow-wave system in a RB-TWT allows for a very low phase velocity. Consequently, the operating voltage of an RB-TWT can be reduced to the range of thousands or even hundreds of volts. The large cathode emission surface facilitates a significantly higher radial electron beam current compared to that of conventional TWTs, ensuring a larger DC input power capability.

The RB-TWT, proposed by scholars from the former Soviet Union in the 1960s, has garnered considerable attention from researchers due to its ultralow operating voltage. However, the stringent demands of radial synchronization have limited the selection of radial SWSs. To date, only the logarithmic helix and its modified configurations, as illustrated in figure 2.2, have been employed for this purpose.

Figure 2.2(a) is an experimental model of an RB-TWT described in the research literature. A planar logarithmic spiral was employed as the SWS, which had an operating frequency band of 200 MHz to 600 MHz, a gain of 10–22 dB, a maximum output power of 50–100 mW, and an operating voltage of 35–40 V. Figure 2.2(b) shows an RB-TWT designed for use with the C band, which adopted the radial logarithmic periodic ring–rod structure as the SWS. The reference [4] does not provide specific performance parameters; however, it has been reported that the RB-TWT exhibits higher power capabilities and a narrower bandwidth than those of the

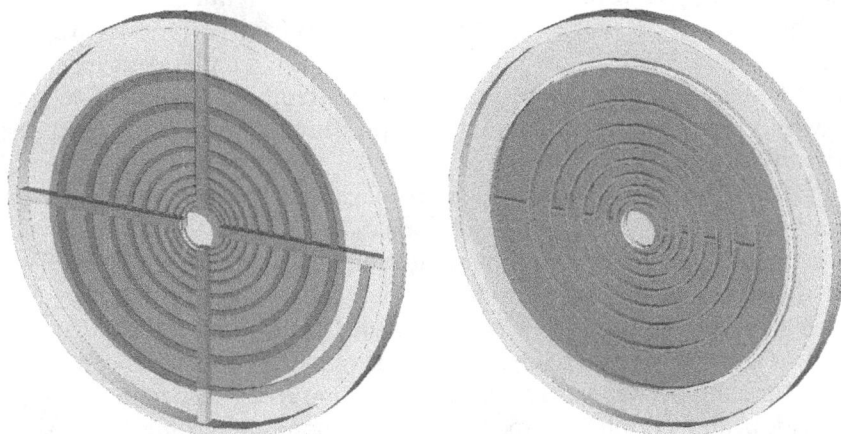

Figure 2.2. Two typical radial beam SWSs. (a) A logarithmic spiral slow-wave system wound with copper wire and (b) a logarithmic periodic loop–rod slow-wave system.

logarithmic TWT. In addition, Norwegian scholars have conducted a detailed study of the structure depicted in figure 2.2(b), extending its operating frequency to the K band. According to computer simulation results, this structure demonstrates a gain of 50 dB, an efficiency ranging between 35% and 40%, and an operational voltage of 300 V [5].

Compared with conventional TWTs, RB-TWTs exhibit ultralow operating voltage along with relatively higher gain and efficiency. Consequently, it is important to conduct research into RB-TWTs, as it offers the potential to broaden their application scope and enhance their competitiveness, particularly in low-frequency bands [5].

2.1.2 Angular radial meander-line SWS

The substantial challenge in increasing the output power of the RB-TWT lies in its very low operating voltage. As a result of the low voltage used, the RB-TWT typically only reaches the milliwatt level. Attempting to enhance the operating voltage by increasing the distance between two adjacent turns of the SWS is met with the drawback of significantly enlarging the radial size of the SWS. This, in turn, causes a reduction in the operating frequency. Consequently, achieving a simultaneous improvement in operating voltage and a reduction in structural size for planar logarithmic spiral TWTs becomes difficult. This limitation severely restricts the potential application fields of such ultralow-voltage TWTs. Considering these challenges, we propose the implementation of a slow-wave system featuring a plane angular logarithmic meander-line SWS [6, 7].

A planar angular logarithmic meander line can be obtained from an ordinary planar logarithmic helix by cutting a sector section from the planar logarithmic helix and connecting the ends of the helix alternately. Compared with the ordinary planar logarithmic SWS, an EMW on the planar angular log-period meander line exhibits a higher phase velocity. This characteristic translates to a higher operating voltage, addressing the challenge of the low output power associated with traditional planar logarithmic SWSs. However, the operating voltage of hundreds of volts is still considerably lower than that of the helical, coupled cavity, folded waveguide, and other conventional TWTs. In addition, the latter are smaller and have a shorter transverse length, enabling operation in higher frequency bands. The TWT can operate with an angular radial banded electron beam. Alternatively, multiple slow-wave circuits can be integrated onto a toroidal substrate to share a radial sheet electron beam. This approach is easier to focus and has the potential to achieve higher output power.

In this section, we commence with an introduction to the design of an SWS which features a plane angular log-period meander line. Subsequently, we derive the dispersion equation for the slow-wave system. Following that, we explore the microstrip angular log-period meander-line slow-wave system in depth. Through computer simulation, we extract information about its dispersion, transmission characteristics, and coupling impedance. Leveraging electromagnetic simulation software, we analyze the impact of various structural parameters on the high-frequency characteristics. Finally, employing the particle-in-cell (PIC) method, we study nonlinear beam interaction. As a result of these comprehensive investigations, we design Ka-band angular log-zigzag slow-wave system TWTs with operating voltages of 1642 V and 809 V.

2.1.2.1 Model and design

The plane angular log-period meander-line SWS represents an enhancement over the conventional planar logarithmic spiral (figure (2.2)). The logarithmic periodic zigzag line at the plane angle is obtained by cutting a specific angle upward from the plane logarithmic helix and alternately connecting the endpoints of the cut section with radial line segments, as depicted in figure 2.3(b).

In contrast to the plane logarithmic helix, the angular logarithmic meander line introduces only one additional parameter, namely the angle θ. A comparison between figures 2.3(a) and (b) reveals that the dimensional ratio of the planar angular log-period meander line is greater than that of the plane logarithmic spiral line. This increase in the dimensional ratio is expected to enhance the operating voltage of the planar angular log-period meander line.

2.1.2.2 Principle of the planar angular log-period meander-line SWS

As in the research method employed for the planar logarithmic spiral, we assume that a plane EMW with phase velocity v_s propagates along the trajectory of the planar angular log-period meander line, as shown in figure 2.4. The trajectory can take various forms, such as a single wire, a double wire, a waveguide, etc. In this context, the corresponding radial EMW on the surface of the meander line exhibits a phase velocity v_r.

When the EMW propagates along the trajectory ABCDE, the radial surface wave propagation time should equal that of the planar EMW propagation time, or the difference in the oscillation period $T = 2\pi/\omega$ must be an integer multiple. Consequently, the relationship can be expressed as follows:

$$\frac{l_{\overset{\frown}{AB}} + l_{BC} + l_{\overset{\frown}{CD}} + l_{DE}}{v_s} = \frac{d_{AE}}{v_r} + nT, \quad n = 0, \pm 1, \pm 2... \tag{2.1}$$

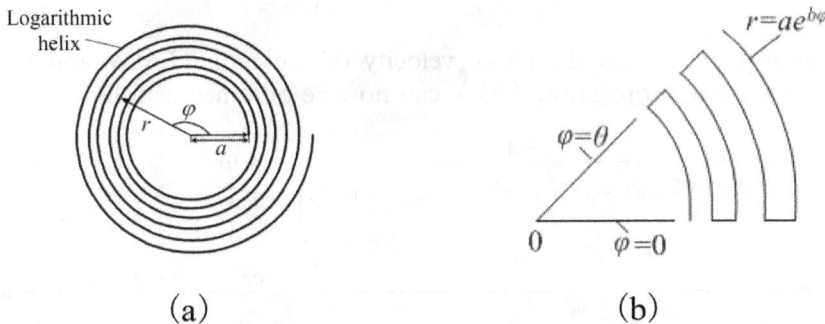

Figure 2.3. Comparison of the (a) planar logarithmic helix SWS and (b) the angular log-period meander-line SWS. © 2013 IEEE. Reprinted, with permission, from [6].

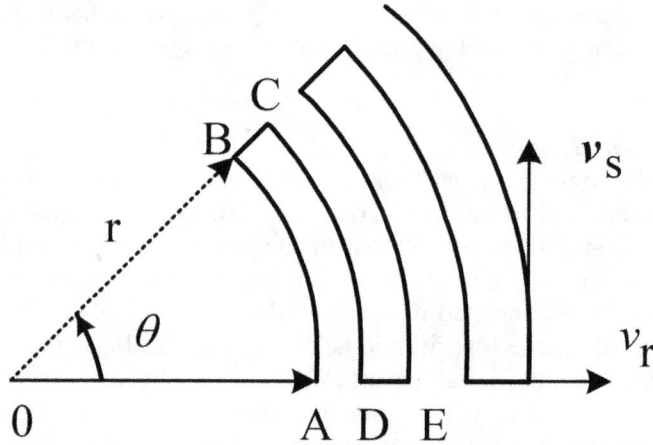

Figure 2.4. Propagation trajectory of an EMW on a planar angular log-period meander line. © 2013 IEEE. Reprinted, with permission, from [6].

where l_{AB} and l_{CD} are the lengths of arcs AB and CD, respectively; l_{BC}, l_{DE} and d_{AE} are the lengths of line segments BC, DE, and AE, respectively; and φ is the polar angle of point A.

Starting from the equation for a logarithmic spiral, $r = ae^{b\varphi}$, the arc length of each segment can be calculated as follows [6]:

$$l_{AB} = \frac{a\sqrt{b^2 + 1}}{b}e^{b\varphi}(e^{b\theta} - 1) \tag{2.2}$$

$$l_{CD} = \frac{a\sqrt{b^2 + 1}}{b}e^{b\varphi}e^{2b\pi}(e^{b\theta} - 1) \tag{2.3}$$

$$l_{DE} = ae^{b\varphi}e^{2b\pi}(e^{2b\pi} - 1). \tag{2.4}$$

Similarly, the radial distance from point A to E is:

$$d_{AE} = ae^{b\varphi}(e^{4b\pi} - 1). \tag{2.5}$$

The relationship between the phase velocity of the radial EMW and the phase velocity of the line-propagating EMW can now be obtained as follows:

$$\begin{cases} \dfrac{v_r}{v_s} = \dfrac{e^{4b\pi} - 1}{\left[\frac{\sqrt{b^2+1}}{b}(e^{b\theta} - 1)(1 + e^{2\pi}) + (e^{2b\pi} - 1)(e^{b\theta} + e^{2b\pi})\right]} \quad (n = 0) \\[4ex] \dfrac{ae^{b\varphi}\left[\frac{\sqrt{b^2+1}}{b}(e^{b\theta} - 1)(1 + e^{2b\pi}) + (e^{2b\pi} - 1)(e^{b\theta} + e^{2b\pi})\right]}{v_s} = \dfrac{ae^{b\varphi}(e^{4b\pi} - 1)}{v_r} + nT(n \neq 0). \end{cases} \tag{2.6}$$

It is evident that for the fundamental mode (n = 0), the ratio of v_r to v_s is independent of a and φ; that is, if b and θ are given, the ratio of the radial phase velocity to the tangential phase velocity of the log-period meander line can also be determined. In general, for a single-frequency plane wave, the propagation velocity in a uniform medium remains constant. Consequently, the radial EMW phase velocity should also remain constant. This constancy implies that the fundamental mode of the EMW on the angular meander line can continuously interact with the radial electron beam, amplifying the signal. For harmonic modes (n ≠ 0), there is no straightforward dependence between v_r and v_s. The radial phase velocity depends on the parameters a and φ. Specifically, the radial phase velocity increases with an increase in the radius. Consequently, the EMW associated with these modes cannot continuously interact with the radial electron beam, precluding its ability to undergo amplification.

We assume that the planar log-period meander line is constructed from an ideal transmission line and that the EMW phase velocity is equal to the speed of light c. Using formula (2.6), we can calculate the normalized radial phase velocity v_r/c for different values of b and θ. The numerical results are shown in figure 2.5.

Therefore, by choosing different values for b and θ, the normalized phase velocity of a planar log-period meander line can be designed to be as low as 0.002, corresponding to an operating voltage of about 1 V, or as high as 0.2, corresponding to an operating voltage of more than 10 537 V. In practical applications, different values of b and θ can be selected based on different requirements.

For a given operating voltage, the specific methods used to calculate b and θ are as follows, taking an operating voltage of approximately 1600 V as an example. First, calculate the corresponding normalized phase velocity according to the given operating voltage, which is about 0.08, and draw the isophase velocity line passing through this point in the figure. Second, determine the corresponding values of (b, θ) by identifying the intersection point of the isophase velocity line and the normalized

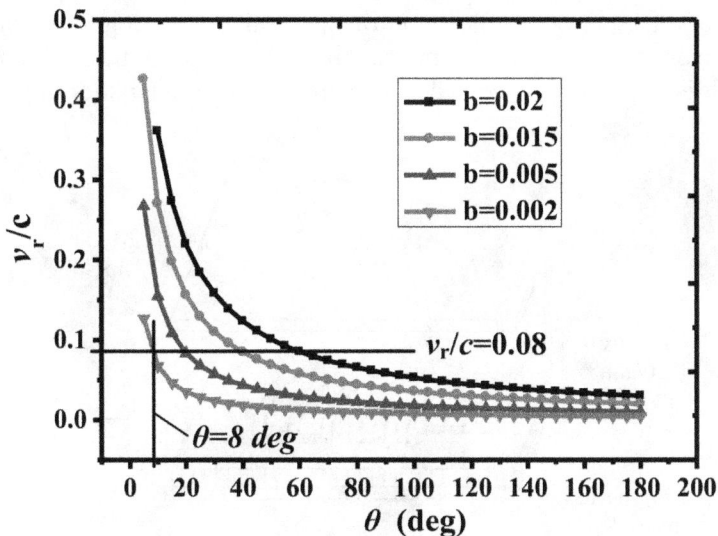

Figure 2.5. Normalized phase velocity curves for different values of b and θ. © 2013 IEEE. Reprinted, with permission, from [6].

phase velocity curve. Multiple sets of (b, θ) values correspond to each normalized phase velocity. In this case, (0.002, 8) are selected because smaller values of b result in a shorter radial length for the corresponding SWS, facilitating the miniaturization of the TWT and making it easier to focus the electron beam.

2.1.2.3 Microstrip angular log-period meander-line SWS

In practical applications, the SWS of the planar log-period meander line is typically implemented using waveguides, ridge waveguides, microstrip lines, or coplanar waveguides. In this section, we will examine the high-frequency characteristics of the microstrip-type planar log-period meander-line slow-wave system. Microstrip lines offer advantages such as smaller size and lower processing costs compared to other transmission lines, making them a promising choice for various applications.

Figure 2.6 shows an SWS model of a microstrip planar angular log-period meander line, which includes a fan-shaped dielectric plate and a planar angular log-period meander line printed on the dielectric plate. The whole structure is enclosed in a fan-shaped metal cavity, and the cavity between the dielectric plate and the upper metal shield can be used as an electron beam tunnel. Such an SWS may operate with a fan-shaped electron beam, which is transmitted from one end of the SWS and moves radially to the other end.

It is evident that the main structural parameters of the slow-wave system of the planar angular log-period meander line are as follows: the initial radius of the microstrip line a, the width w, the thickness t, the angle θ of the SWS, the period coefficient b, the thickness of the dielectric plate d, and the distance between the microstrip line and the upper metal shield h.

2.1.2.3.1 Dispersion characteristics

We have previously examined the relationship between the normalized phase velocity, the angle θ, and the period coefficient b for the ideal angular log-period meander-line SWS. However, since the actual microstrip behaves as a transmission line with dispersion, it is imperative to consider its dispersion characteristics in our analysis.

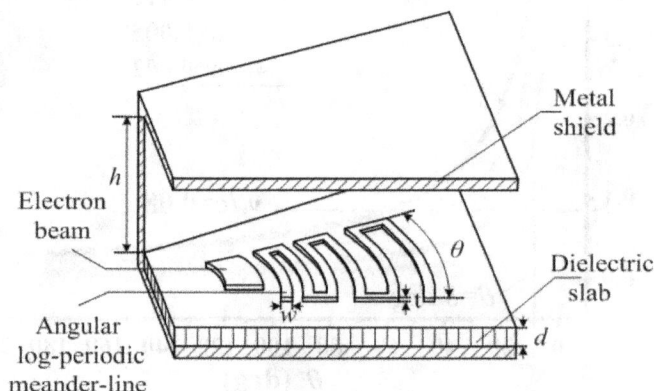

Figure 2.6. Schematic diagram of a microstrip angular log-period meander-line SWS. © 2013 IEEE. Reprinted, with permission, from [7].

Given that any change in the structural parameters affects the transmission characteristics of the slow-wave system, our current focus is solely on the influence of changes in the angle θ and the period coefficient b of the SWS on the normalized phase velocity.

Figures 2.7(a) and (b) show how the normalized phase velocity of the slow-wave system changes for different values of the period coefficient b for θ values of 5° and 10°, respectively.

It can be observed that, with a fixed angle θ for the SWS, the normalized phase velocity of the SWS increases as the period coefficient b increases. When θ and b are fixed, the normalized phase velocity of the SWS exhibits minimal change over a broad frequency band, indicating that the structure possesses a wide operating frequency range. In addition, with an increase in b, the curvature of the dispersion curve at the high-frequency end becomes obvious, indicating a strengthening of dispersion. It is important to note that, for computational efficiency, the chosen number of periods for the slow-wave system in this analysis is ten. Moreover, the influence of transmission characteristics is not considered here. Although the transmission at the high-frequency end may be suboptimal, it is still included in the figure.

Figures 2.8(a) and (b) show how the normalized phase velocity of the SWS changes with the angle θ for periodic coefficient b values of 0.01 and 0.005, respectively. It can be observed that the normalized phase velocity of the microstrip planar angular log-period meander-line SWS decreases with an increase in the angle θ, and the frequency range corresponding to the dispersion curve becomes progressively smaller. This trend is explained by considering the transmission characteristics used in the calculation process. As the angle θ increases, the transmission characteristics of the slow-wave system markedly deteriorate at the high-frequency end.

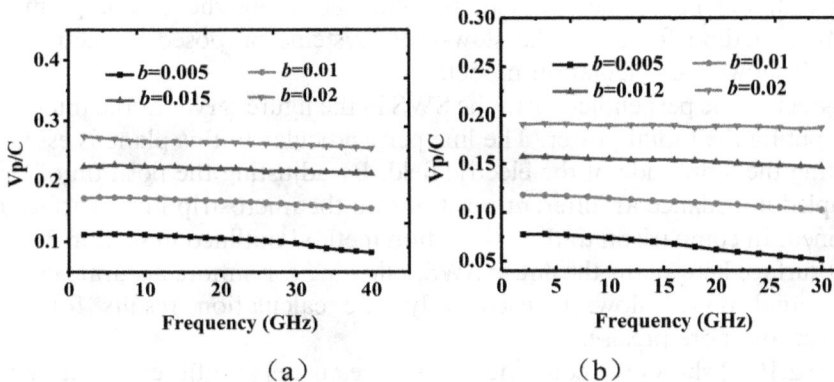

Figure 2.7. Variations in the dispersion characteristic curves (presented in the form of normalized phase velocity versus frequency) for different values of b and fixed values of θ. (a) $\theta = 5°$; (b) $\theta = 10°$. © 2013 IEEE. Reprinted, with permission, from [7].

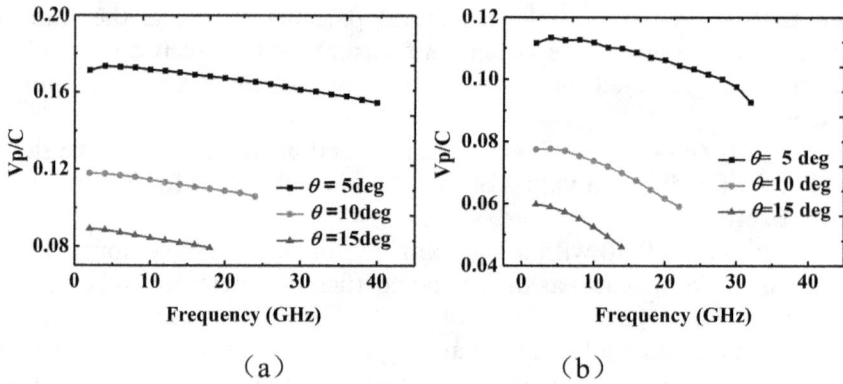

Figure 2.8. Variations in the dispersion characteristic curves (presented in the form of normalized phase velocity versus frequency) for different values of θ and fixed values of periodic coefficient b. (a) $b = 0.01$; (b) $b = 0.005$. © 2013 IEEE. Reprinted, with permission, from [7].

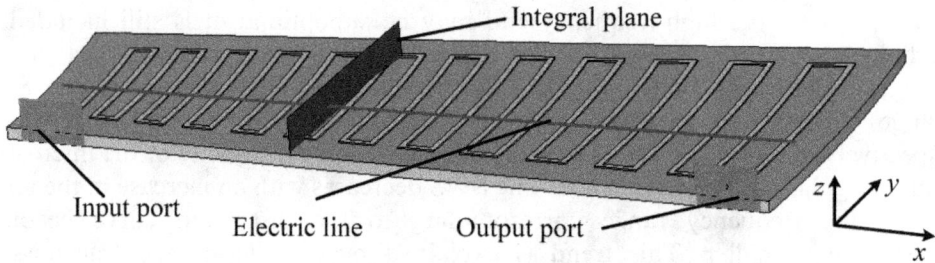

Figure 2.9. Coupling impedance calculation model used for microstrip planar angular log-period meander-line SWSs.

2.1.2.3.2 *Coupling impedance*

The microstrip planar angular log-period meander-line SWS is an aperiodic SWS, and its coupling impedance K_c can be obtained using the coupling impedance calculation method for aperiodic slow-wave systems proposed in section 3.2.2.2. Figure 2.9 shows the calculation model.

The green plane perpendicular to the SWS in the figure serves as the integral plane for computing the radial power. The line perpendicular to this plane is essential for calculating the amplitude of the electric field. By adjusting the position of this line, the coupled impedance at different positions on the microstrip line's surface can be determined. In comparison to the calculation method outlined in section 3.2.2.2, the integral surface here spans the entire SWS, allowing for a more accurate calculation of the radial power flow. Consequently, the calculation results for coupling impedance are more precise.

Figure 2.10(a) shows the coupling impedance curves at different positions on the surface of a slow-wave system with a microstrip planar angular log-period meander-line SWS at angles of $\varphi = 0°$, $5°$, and $10°$.

As the distance to the microstrip line increases, the coupling impedance of the planar angular log-period meander-line SWS decreases rapidly. For instance, when the distance is 0.015 mm, the maximum coupling impedance can exceed 90 Ω, while at 0.1 mm, the coupling impedance decreases to 0.02 Ω. Figure 2.10(b) shows the variation of coupling impedance at 0.04 mm from the surface of the microstrip line as the angle from the x-axis increases. It can be seen that the coupling impedance is slightly larger near the edge of the microstrip planar angular log-period meander-line SWS. The lowest values are observed at the central angle, primarily due to the larger radial component of the electric field at the edge.

In this section, we extend our analysis to consider the coupling impedance of the slow-wave system of a microstrip planar angular log-period meander-line SWS with different angles θ and different periodic coefficients b. Figure 2.11 shows the variation of coupling impedance with frequency for different values of θ and b. The coupling impedance here is the average in-plane coupling impedance from 0.04 mm to 0.06 mm from the surface of the microstrip. Figure 2.11(a) shows the

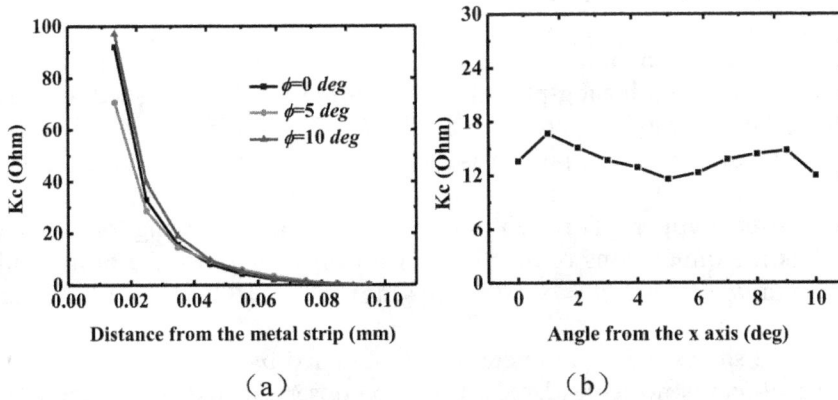

Figure 2.10. Coupling impedance at different positions on the surface of the microstrip planar angular log-period meander-line SWS. (a) Curve showing the variation of the coupling impedance versus the distance to the surface of the microstrip line. (b) Curve showing the variation of the coupling impedance versus the angle from the x-axis.

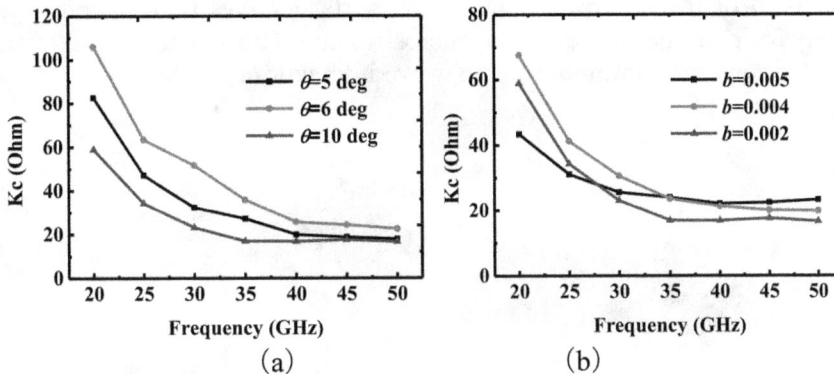

Figure 2.11. Influence of the main structural parameters on the coupling impedance. (a) Variation for different values of θ; (b) variation for different values of b.

change of coupling impedance with θ when $b = 0.002$. It can be seen that the change in coupling impedance with θ is not monotonous but first increases and then decreases. Figure 2.11(b) shows the change of coupling impedance with b when $\theta = 10°$. It can be seen that the coupling impedance changes irregularly with b below 35 GHz but increases with an increase in b above 35 GHz.

2.1.2.3.3 Transmission characteristics

The key parameters determining the dispersion characteristics of the slow-wave system are the period coefficient b and the angle θ. These parameters can be determined based on the operating voltage requirements using the method described earlier. Research results show that both b and θ exert an important influence on the transmission characteristics of the SWS. In addition to b and θ, the initial radius a, the strip width w, the strip thickness t, and the dielectric slab thickness d also contribute to the transmission characteristics. A comprehensive consideration of these parameters is necessary to achieve improved transmission characteristics.

Figure 2.12 shows a simulation model of the transmission characteristics of a microstrip planar angular log-period meander-line SWS established in Computer Simulation Technology's Microwave Studio (CST's MWS). Here, 30 periods are adopted, coaxial lines are employed as the output structure, the dielectric slab is set to boron nitride, and the metal strip and the inner conductor of the coaxial line are all oxygen-free copper. Through optimization, the following parameters are obtained as the dimensions of an SWS for an operating voltage of around 1600 V: $a = 3$ mm, $b = 0.002$, $\theta = 8°$, $d = 0.04$ mm, $h = 0.16$ mm, $w = 0.02$ mm, and $t = 0.005$ mm.

Figure 2.13 shows the S-parameter curve obtained by simulation (the loss of the slow-wave system is not considered). It can be observed that within the frequency range from 0 to 40 GHz, the transmission coefficient S_{21} is greater than -0.1 dB and the reflection coefficient S_{11} is less than -22.5 dB, indicating favorable transmission characteristics.

Figure 2.14(a) shows the horizontal electric field, and figure 2.14(b) shows the vertical field. Observation indicates that when an EMW propagates along the meander path of the microstrip line, radial surface waves that exhibit noticeable periodicity form on the surface of the microstrip line. This phenomenon satisfies one of the conditions for continuous beam–wave interaction.

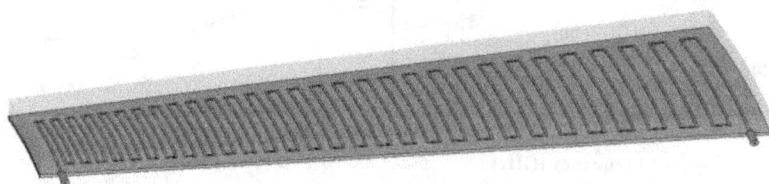

Figure 2.12. Microstrip planar angular log-period meander-line SWS showing its input/output structure. © 2013 IEEE. Reprinted, with permission, from [7].

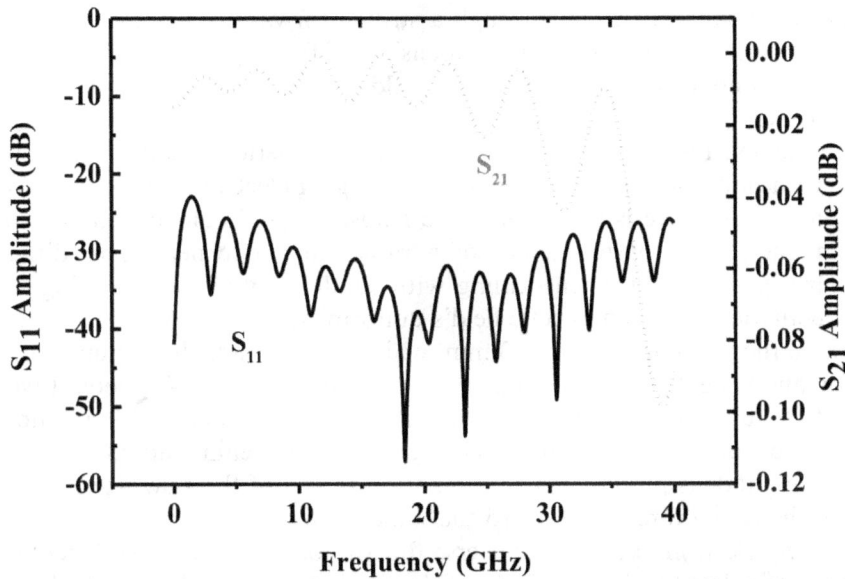

Figure 2.13. Simulated *S*-parameters. © 2013 IEEE. Reprinted, with permission, from [6].

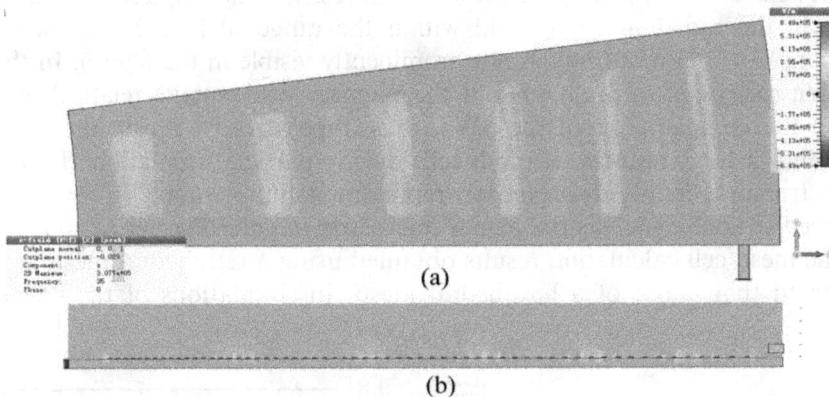

Figure 2.14. Electric field diagram for the microstrip planar angular log-period meander-line SWS. (a) Horizontal cross-sectional distribution. (b) Vertical cross-sectional distribution. © 2013 IEEE. Reprinted, with permission, from [6].

2.1.2.3.4 Beam–wave interaction

As mentioned earlier, the microstrip planar angular log-period meander-line SWS is created by cutting and deforming a fan-shaped segment of the planar logarithmic spiral SWS. Consequently, this type of SWS can also operate by cutting a portion from a radially diverging or converging sheet electron beam at the same angle. For the convenience of study, a simulation method involving a complete radial sheet electron beam is directly employed here.

In the particle simulation of beam interaction, the electron gun can be simplified to an ideal emission surface, allowing for the generation of an electron beam with any desired shape and current through a straightforward setup. Consequently, the primary concern revolves around the focusing of the angular radial sheet electron beam. To address this, a radial magnetic field can be employed to focus the radial electron beam.

Theoretically, the microstrip planar angular log-period meander-line SWS can operate with both radially diverging and converging electron beams. However, due to the passive characteristics of magnetic fields, it is difficult to obtain a uniform radial magnetic field at the center of a radial magnetic field. This difficulty is particularly pronounced when dealing with smaller initial radii, leading to more severe fluctuations in the magnetic field's uniformity. The initial radius of the SWS obtained in the previous section is 3 mm, which is just inside the nonuniform region of the magnetic field. If the electron beam were to enter the SWS here, it would be difficult to focus. In order to minimize the influence of magnetic field inhomogeneity on electron beam focusing, a radial convergent electron beam is adopted here. In this setup, the cathode is placed at the termination radius of the slow-wave system and moves in the radial direction toward the initial radius.

For given electron beam parameters, the Brillouin magnetic field required for focusing is calculated to be 0.52 T. Considering that the boundary of the fan-shaped RSEB is different from that of the complete RSEB, 0.65 T is chosen as the value of the magnetic field used for focusing, as shown in figure 2.15. The graph shows a noticeable increase in the radial magnetic field within the range of $r < 2$ mm, with slight fluctuations at $r = 3$ mm (though not prominently visible in the figure). In the main interaction zone of 3.5 mm $< r < 8$ mm, the magnetic field remains relatively constant.

Given the compact size of the SWS, it is imperative to divide the simulation domain into a large number of mesh cells during particle simulation. This ensures that the structural details are accurately represented in the simulation results. In fact, in a curved structure such as the planar microstrip angular log-period meander-line SWS, the mesh cell calculation results obtained using a tetrahedral mesh are closer to the truth than those of a hexahedral mesh. In simulations of the beam–wave

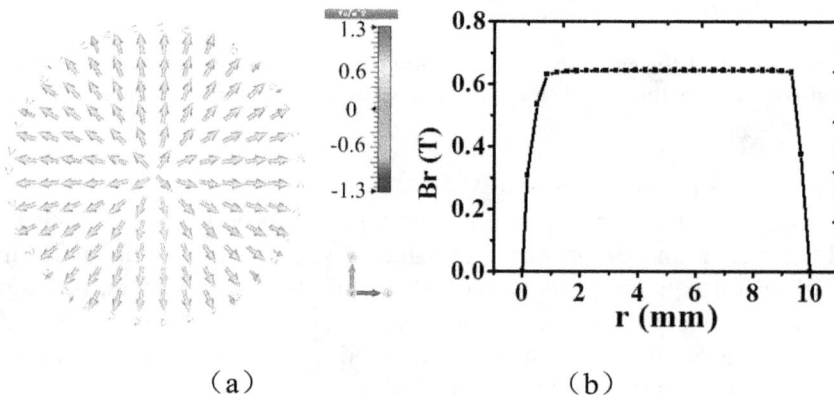

(a)　　　　　　　　　　　　　　(b)

Figure 2.15. Radial magnetic field used in the particle simulation. (a) The field's vector diagram and (b) amplitude curve. © 2013 IEEE. Reprinted, with permission, from [6].

interaction, only a hexahedral mesh can be employed. Consequently, during the simulation process, it is essential to optimize the mesh settings for hexahedra to the greatest extent possible within CST MWS. The transmission's characteristic curve is close to that of a tetrahedral mesh; thus, the same mesh setting is used for the particle simulation. In the particle simulation described in this section, the mesh settings are as follows: the number of grid lines per wavelength is 30, the maximum–minimum grid ratio is 30, and the total number of grids is 7,239,960.

In the simulation, boron nitride is selected as the dielectric material, pure copper is selected for the metal strip and background material, and the distance between the electron beam and the surface of the microstrip line is set to 0.02 mm. The calculated performance data of the microstrip planar angular log-period meander-line SWS are shown in figure 2.16, in which figure 2.16(a) is an electron bunching diagram, figure 2.16(b) shows the variation of output power and gain with frequency, and figure 2.16(c) displays the variation of efficiency with frequency.

It can be observed that a single microstrip planar angular log-period meander-line SWS can generate a peak power that exceeds 100 W within the frequency range of 20 to 42 GHz. The maximum power achieved is 160 W, and the maximum efficiency reaches 19.2%. Notably, the gain curve remains relatively constant in the frequency range of 20 to 45 GHz. These results indicate a large operating bandwidth, and the outcomes of the interaction align with the preceding analysis.

To compare the above device with a traditional helical TWT, we take the example of Netcomsec's LD7338 TWT. In the frequency range of 27.5–31 GHz, the output power of the latter can exceed 517 W and its gain can reach 38.1 dB. The microstrip planar angular log-period meander-line SWS still needs to be improved in terms of output power and gain. Potential improvement methods include increasing the interaction period and incorporating attenuators. Despite the need for refinement, the microstrip planar angular log-period meander-line SWS offers clear advantages, such as operating at only one-tenth of the voltage required by the LD7338, a greater bandwidth, and a significantly smaller size compared to the latter.

2.1.3 Modified angular log-period meander-line SWS

A novel modified angular log-periodic folded waveguide SWS (MALPFW-SWS) for high-gain TWTs is proposed for a single-section high-gain TWT [8].

2.1.3.1 Equation and model
The proposed modified angular log-periodic (MALP) meander line can be obtained using the same evolution method to process a modified logarithm helix whose path can be described by the equation:

$$r = a_0 e^{b_0 \varphi} - r_0 \qquad (2.7)$$

where r and φ are the coordinates in the radial and angular axes, respectively; r_0 is the additional value in the radial axis.

Figure 2.17 shows sketches of the MALPFW-SWS. The dashed line represents the meander path of the MLPFW-SWS, i.e. the MALP meander line. Here, d_r and L

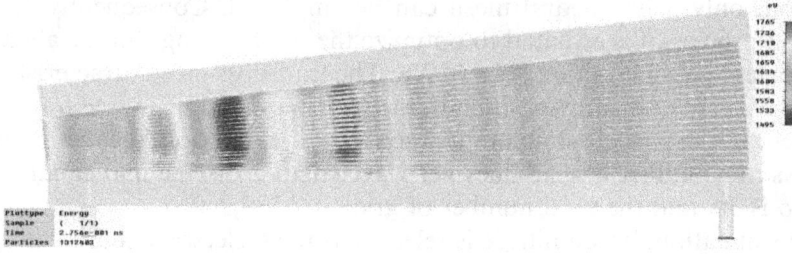

(a)

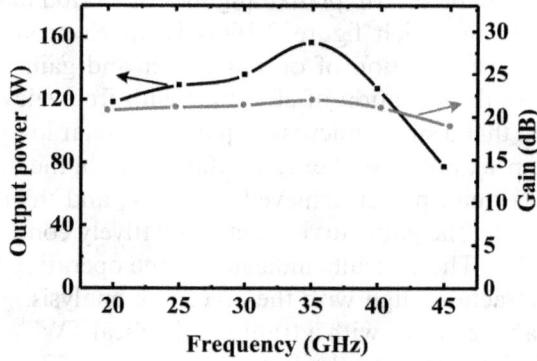

(b)

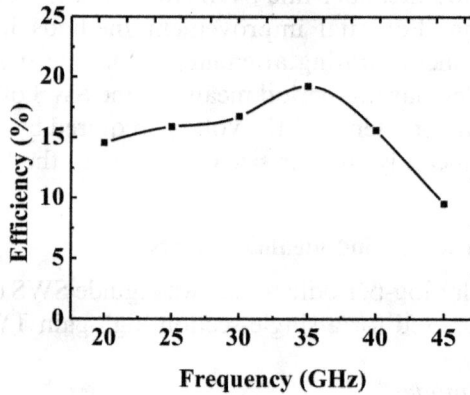

(c)

Figure 2.16. Microstrip planar angular log-period meander-line SWS interaction diagram and performance curves. (a) Electron bunching diagram; (b) output power and gain–frequency curves; (c) efficiency–frequency curve. © 2013 IEEE. Reprinted, with permission, from [6].

are the beam and wave path lengths of a unit, respectively; θ is the angle of the MALP meander line; a and b are the transverse dimensions of the waveguide, and r_c is the radius of the beam hole.

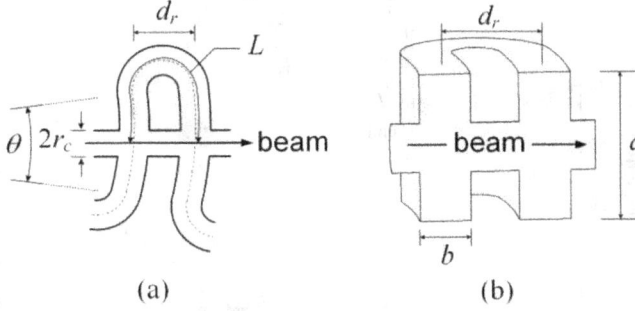

Figure 2.17. Sketch of the novel MALPFW-SWS. (a) Two-dimensional view of one unit on the r–φ plane; (b) 3D model of one unit. © [2020] IEEE. Reprinted, with permission, from [8].

According to the path equation (2.7), the beam and wave path lengths in the nth unit, d_{rn} and L_n, can be written as

$$d_{rn} = a_0(e^{2b_0\pi} - 1)e^{2b_0(n-1)\pi}$$
$$= d_{r1}e^{2b_0(n-1)\pi} \tag{2.8}$$

$$L_n = \frac{a_0}{b_0}\left(e^{\frac{b_0\theta}{2}} - 1\right)(1 + e^{2b_0\pi})e^{2b_0(n-1)\pi} + \frac{d_{rn}\pi}{2} - r_0\theta$$
$$= (L_1 + r_0\theta)e^{2b_0(n-1)\pi} - r_0\theta. \tag{2.9}$$

A simplified dispersion equation for the TE_{10} mode was derived using a smooth-wall folded waveguide SWS (FW-SWS) model, namely:

$$\omega = c\sqrt{\left(\frac{\beta_{zm}d_r - \pi - 2m\pi}{L}\right)^2 + \left(\frac{\pi}{a}\right)^2} \tag{2.10}$$

where ω is the angular frequency of the wave, β_{zm} is the longitudinal propagation constant (rad/m) of the mth spatial harmonic, and c is the speed of light in vacuum.

Upon substituting d_r and L into (2.10), we can find a common intersection point P for these dispersion curves, as shown in figure 2.18. We therefore call it the perfect synchronized point (PSP); its coordinates are

$$\left(\frac{\pi(L_1 + r_0\theta)}{r_0\theta d_{r1}}, \; c\sqrt{\left(\frac{\pi}{r_0\theta}\right)^2 + \left(\frac{\pi}{a}\right)^2}\right) \tag{2.11}$$

in terms of the dimensions of r_0, d_{r1}, L_1, and θ.

For an EMW with a specified frequency, the phase velocity varies with n, somewhat like the behavior of conventional phase velocity tapered SWSs. When the specified frequency is lower than that of the PSP, the phase velocity is positively tapered; it changes to negatively tapered once the frequency is higher than that of the PSP, as shown in figure 2.19(a). Figure 2.19(b) gives the relation between the phase velocity and the axial distance at 31.38 GHz, 34.01 GHz, and 35.44 GHz.

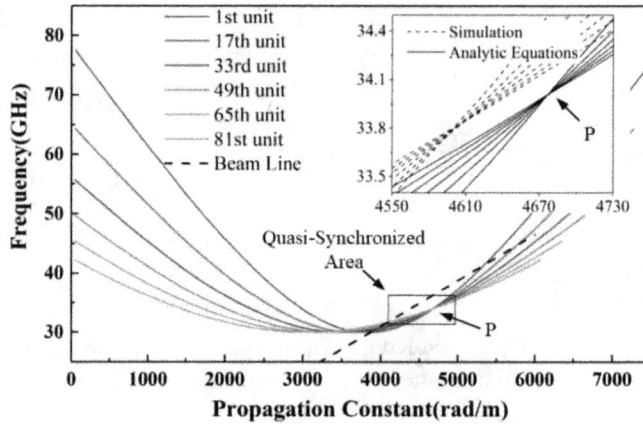

Figure 2.18. Dispersion curves of the fundamental mode of the MALPFW-SWS units. © [2020] IEEE. Reprinted, with permission, from [8].

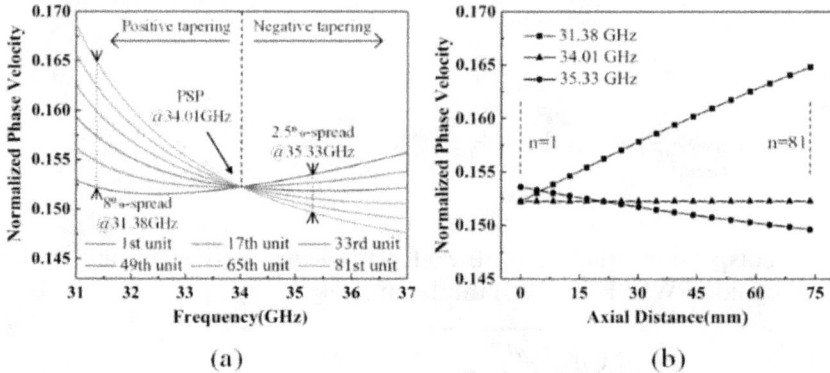

Figure 2.19. Variations of phase velocity with (a) frequency and (b) axial distance. © [2020] IEEE. Reprinted, with permission, from [8].

The quasi-synchronized area (QSA) is the effective interaction area of the MALPFW-SWS. The frequency range of the QSA is defined in two different ways for its two ends. For the lower-frequency end, the phase velocity is positively tapered. The average spread of the positive tapering is about 8%, which was adopted as the spread limit of the lower-frequency end here.

In general, appropriate positive tapering can improve the gain of the TWT. Therefore, the point of top gain should have a lower frequency than that of the PSP in the MALPFW-TWT. As a result, the conservative value of 2.5% is adopted as the spread limit of the upper end. This limitation is valuable when discussing the saturation output power, and it can provide a preliminary guide for the design of the MALPFW-SWS.

If the input frequency falls outside the frequency range of the QSA, the TWT gain decreases rapidly due to the abrupt variation in phase velocity. So, even if the beam line crosses the 'oscillation area' (such as the band edges shown in figure 2.18), the oscillation is suppressed because the oscillation frequency falls outside the QSA range.

In addition, the simplified dispersion equation (equation (2.10)) deviates from experimental measurements by about 1%–2% because the effects of circuit bends and beam holes were not considered. If needed, more accurate dispersion curves could be obtained using simulation methods. The difference between the simulation results and the analytic equations is also shown in figure 2.18.

2.1.3.2 Design and simulation results

In order to validate the theory, a Ka-band MALPFW-SWS was designed and studied by PIC simulation. Table 2.1 shows the projected dimensions of the designed SWS. N is the total number of units used in the SWS. According to the analytic expressions presented in section 2.1.3.1, the normalized phase velocity of the PSP is 0.152, corresponding to an optimized operating beam voltage of 6.9 kV.

An ideal round electron emitter with a radius of 0.27 mm was selected in the PIC simulation to provide a round electron beam at a current of 300 mA. A solenoidal magnetic field of 0.4 T was applied to focus the electron beam.

The output power was calculated to be 217 W when the input power is set to 18 mW at the frequency point of 32.4 GHz, as shown in figure 2.20. The output signal was very stable over the simulation duration of 50 ns, and the gain of the one-section MALPFW-TWT even reached 40.8 dB. The corresponding frequency spectrums shown in figure 2.20 also prove that oscillations have been suppressed.

Table 2.1. Projected dimensions of the designed Ka band MALPFW-SWS.

Symbols	a_0	b_0	r_0	θ
Values	290 mm	0.000 45	268 mm	2°
Symbols	N	a	b	r_c
Values	81	5 mm	0.45 mm	0.34 mm

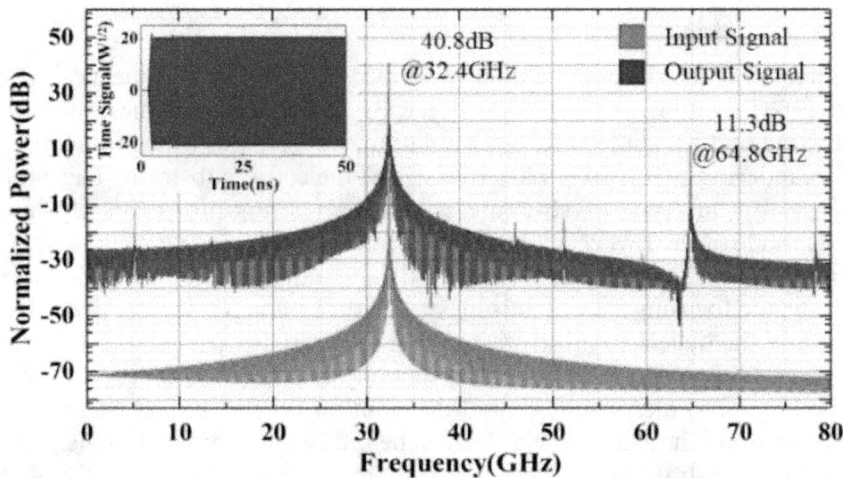

Figure 2.20. Frequency spectrums of the input and output signals (saturated). © [2020] IEEE. Reprinted, with permission, from [8].

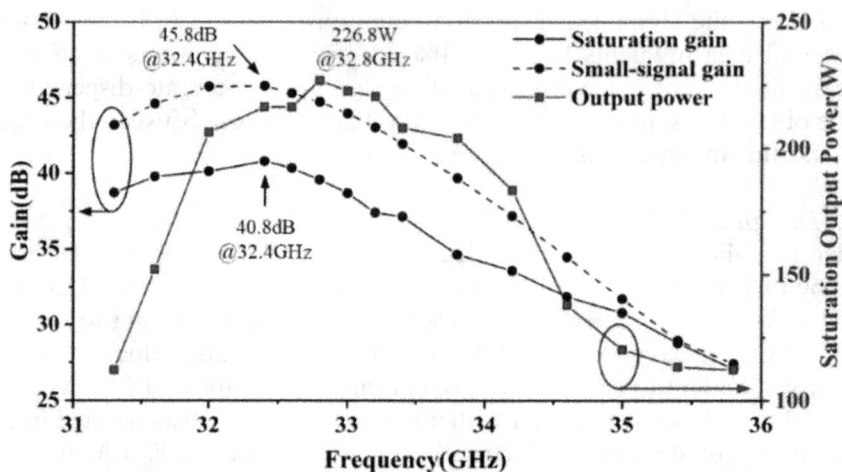

Figure 2.21. Output power (saturated at 32.8 GHz), saturation gain, and small-signal gain versus frequency. © [2020] IEEE. Reprinted, with permission, from [8].

Figure 2.21 shows the variations of the saturation input/output powers and the saturation/small-signal gains versus frequency.

We find that the 3 dB frequency band of the saturation output power is 31.3–35.8 GHz from the blue line shown in figure 2.21. Furthermore, the small-signal gain obtained by fixing the input power at 50 µW and sweeping the frequency indicates that the 3 dB frequency band of the small-signal gain is 31.3–33.2 GHz. The maximal small-signal gain is 45.8 dB, which occurs at 32.4 GHz; this frequency is lower than that of the PSP, as predicted.

2.2 Axial periodic slow-wave structures

2.2.1 Meander strip-line slow-wave structures

The primary characteristic of the microstrip meander-line SWS is dielectric substrate support [9]. The fabrication process typically includes techniques such as mask exposure, magnetron sputtering, laser ablation, and others, which are employed to create meandering metal strip patterns on the dielectric substrate. The processing technology for this type of structure is relatively straightforward. However, the operating mode of the EMW is a quasi-surface wave, demanding a high-energy sheet electron beam to skim over the surface of the meander line at a close distance. Due to the magnetic focusing system, it is inevitable that some electrons will bombard the strip line and the dielectric substrate, causing the strip line to fuse and accumulating charge in the dielectric. The former can directly result in the failure of the device, while the latter can alter the potential distribution on the surface of the SWS and further deteriorate the shape of the electron beam. In addition, due to the presence of the dielectric substrate, the electric field is more concentrated in the dielectric, resulting in a lower coupling impedance in the beam–wave interaction region of the surface.

To address the shortcomings of the traditional planar SWS made of microstrip lines, a planar SWS made from suspended strip metal lines is proposed, which operates in the Ka band due to dielectric rods clamped to both sides. A thicker metal strip can be created by laser cutting a molybdenum sheet. At the same time, the dielectric rods are supported on both sides rather than the bottom, enabling the electron beam to fully interact with the surface wave at a close distance, which helps to avoid the inherent issues such as charge accumulation, device failure, and low coupling impedance of microstrip devices. A detailed analysis of the key operating characteristics of this device, including its dispersion, transmission, and beam–wave interaction, was carried out. This section provides a comprehensive understanding of the basic principles and thoroughly describes the simulation process. The universality of this approach to different SWSs is highlighted, serving as the foundation for subsequent chapters.

2.2.1.1 Model

Figure 2.22 shows the single-period 3D model of the planar strip-line SWS, in which parts 2.22(a)–(c) respectively show the three-dimensional view, the x–y plane, and the x–z plane of the structure. The gray part of the model is the metal shell made of oxygen-free copper (conductivity 5.8×10^7 S m^{-1}), the gray metal meander line is made of molybdenum (conductivity 2×10^7 S m^{-1}), and the yellow parts are dielectric rods of boron nitride (dielectric constant $\varepsilon_r = 4$, loss tan $\delta = 0.0005$).

The plan view of the strip is a 'U' shape, the bottom of which lies on the z-axis; the spacing between the arms of the 'U' is w. The cross-section of the dielectric rod is a 'T' shape. The selection of the convex type is driven by two main considerations. First, it facilitates the attachment and fixation of the metal shell. Second, it helps to minimize the proportion of dielectric material in close proximity to the strip line. Key structural dimensions are also identified in the figure, and the related parameters are shown in table 2.2.

2.2.1.2 Dispersion characteristics

The simulation process used to obtain the dispersion curve is as follows: first, using the eigenmode solver to scan the phase delay φ of a single period, the frequency-phase curve of multiple modes of the period can be obtained. Second, based on the relationship given in equation (2.12) for the phase velocity v_p, the angular frequency ω, and the phase constant β, the dispersion relationship of the structure can be calculated. Formula (2.13) presents a quick calculation formula for the normalized phase speed v_{pn}, where the frequency f_1, the period length p_1, and the phase shift φ_1 are represented in GHz, mm, and ° (in degrees), respectively.

$$v_p = \frac{\omega}{\beta} = \frac{2\pi f p}{\varphi} \tag{2.12}$$

$$v_{pn} = \frac{v_p}{c} = \frac{2\pi f p}{\varphi c} = \frac{1.2 f_1 p_1}{\varphi_1} \tag{2.13}$$

(a)

(b)

(c)

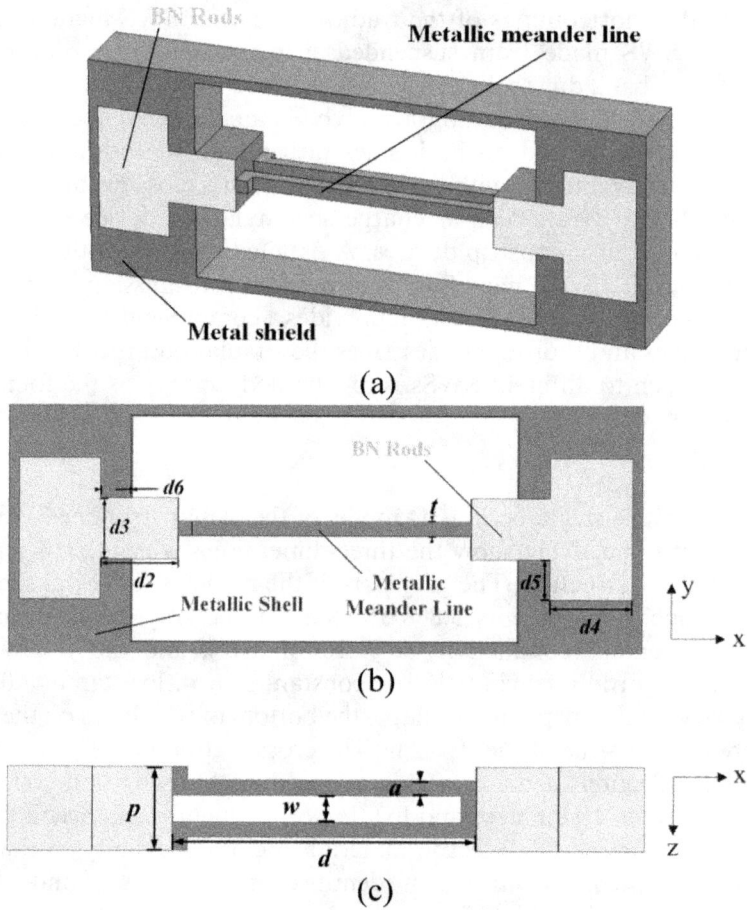

Figure 2.22. Schematic diagrams of the single-period structure of the strip-line planar SWS. (a) Three-dimensional structure; (b) x–y plane; (c) x–z plane. Reprinted from [9], with the permission of AIP Publishing.

Table 2.2. The dimensional parameters of the planar SWS optimization of a strip line held by dielectric rods.

Structural parameter	Value (mm)	Structural parameter	Value (mm)
a	0.07	$d2$	0.38
w	0.13	$d3$	0.3
d	1.3	$d4$	0.4
p	0.4	$d5$	0.2
t	0.07	$d6$	0.15

$$U_0 = \frac{1}{2}\frac{m}{e}v_e^2 \tag{2.14}$$

$$m = \frac{m_0}{\sqrt{1 - \frac{v_e^2}{c^2}}}. \tag{2.15}$$

Once the fixed operating frequency point is selected, the axial velocity v_p of the wave is calculated. We set it equal to the electron velocity v_e, and then the operating voltage of the synchronization point can be calculated. The voltage value is determined by equations (2.14) and (2.15). In the operation of the TWT, where energy is transferred from electrons to an EMW, the operating voltage applied to the electron beam is slightly higher than the voltage calculated from the beam–wave synchronization velocity. This approach differs from the approach used for particle accelerators, where the process is reversed.

Figure 2.23 shows the dispersion curve of the structure and the operating voltage line of the TWT. The intersection occurs at 30 GHz, and within the range of 30–42 GHz, the voltage line is higher than the dispersion line, indicating a potential area for EMW amplification. Below this frequency band, electrons are accelerated, but the output power of the EMW is less than the input power. If the frequency is higher than this range, the energy transfer is hindered because the frequencies of the voltage line and the dispersion line are too far apart to interact synchronously. A quantitative analysis of this aspect can be obtained from a simulation of the particle beam–wave interaction.

2.2.1.3 Coupling impedance

In general, the greater the coupling impedance, the stronger the beam–wave interaction of the TWT, leading to a higher output power. The value of this parameter is defined by equation (2.16). It mainly depends on the amplitude of the longitudinal electric field E_z in the slow-wave circuit and the transmitted power flow P. Equations (2.17) and (2.18) define the axial electric field intensity, and equation (2.19) defines the power flow:

$$K_{cn} = \frac{|E_{zn}|^2}{2\beta_n^2 P} \tag{2.16}$$

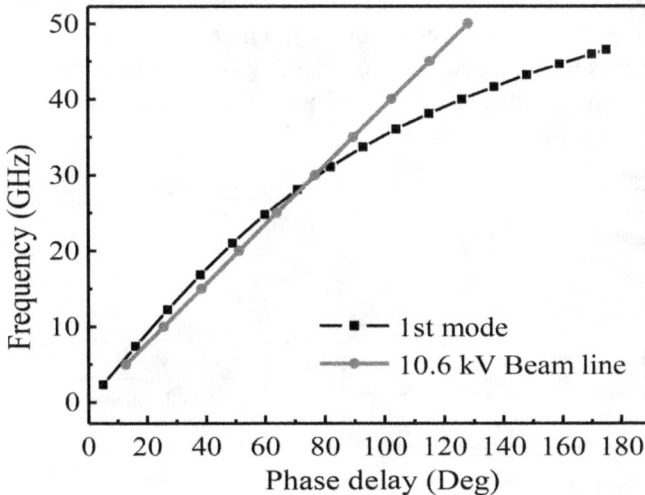

Figure 2.23. Dispersion curve of the single period structure and the voltage line of 10.6 kV. Reprinted from [9], with the permission of AIP Publishing.

$$E_{zn} = \frac{1}{p} \int_0^p E_z(z) \cdot e^{j\beta_n z} dz$$

$$= \frac{1}{p} \int_0^p [\text{Re}(E_z) + j\text{Im}(E_z)] \cdot e^{j\beta_n z} dz$$

$$= \frac{1}{p} \int_0^p [\text{Re}(E_z) \cdot \cos(\beta_n z) - j\text{Im}(E_z)\sin(\beta_n z)] dz$$

$$+ j\frac{1}{p} \int_0^p [\text{Re}(E_z) \cdot \sin(\beta_n z) + j\text{Im}(E_z)\cos(\beta_n z)] dz$$

(2.17)

$$|E_{zn}|^2 = \left\{ \frac{1}{p} \int_0^p [\text{Re}(E_z) \cdot \cos(\beta_n z) - \text{Im}(E_z)\sin(\beta_n z)] dz \right\}^2$$

$$+ \left\{ \frac{1}{p} \int_0^p [\text{Re}(E_z) \cdot \sin(\beta_n z) + \text{Im}(E_z)\cos(\beta_n z)] dz \right\}^2$$

(2.18)

$$P = \int_S \vec{S}_{av} \cdot d\vec{S}.$$

(2.19)

In the simulation calculation process, the coupling impedance of a single calculation represents the value at a certain point. However, the real coupling impedance is the average coupling impedance over the cross-section of the electron beam. It is more meaningful to calculate the average coupling impedance of the planar SWS, considering that the EMW behaves like a surface wave and its amplitude decreases rapidly as it moves away from the surface of the strip line.

Figure 2.24 shows the cross-sectional diagrams of planar microstrip-type and meander strip-line-type SWSs. While selecting more sample points in the cross-section of the electron beam can lead to more accurate calculation results, a balance needs to be struck between calculation time and accuracy. Therefore, nine points labeled A–I are ultimately chosen as the key points for calculating the coupling impedance. According to the principle of symmetry, A, D, and G are identical to C, F, and I; thus, only six

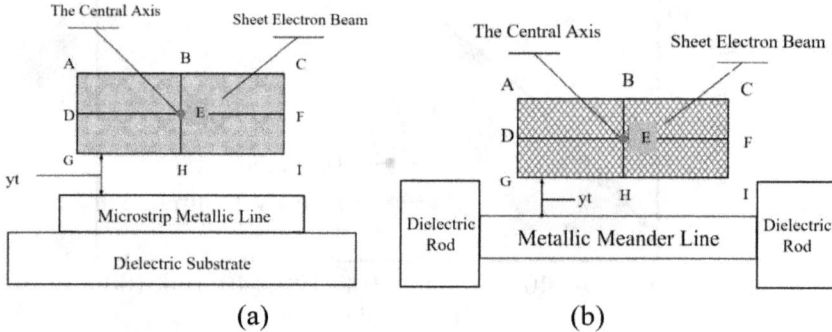

Figure 2.24. Calculations area of the coupling impedance (shadowed area): (a) microstrip structure and (b) strip-line structure.

points—A, B, D, E, G, and H—are calculated. The thickness of the dielectric substrate in the microstrip SWS is 0.2 mm. The other parameters, such as the thickness of the electron beam, the thickness of the strip line, and the distance yt from the bottom of the electron beam to the strip line, are consistent. As observed from figure 2.25, the coupling impedance values of the six sample points in the two structures roughly follow the same variation pattern, with a sequential order of A, B, D, E, G, and H from small to large. In a longitudinal comparison, the coupling impedance increases the closer the sample point is to the surface of the strip line and the closer it is to the middle of the strip line. A transverse comparison shows that the strip line exhibits a more pronounced improvement than that of the microstrip line at the same sampling points.

Figure 2.26 shows the coupling impedance curves of the two structures after averaging i.e., the planar dielectric-rods-support uniform metal meander line

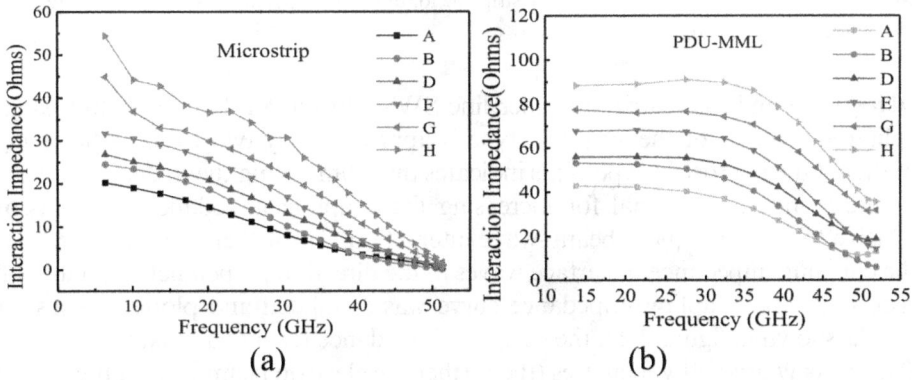

Figure 2.25. Coupling impedance of each point in the interaction region of the two structures: (a) microstrip structure and (b) strip-line structure.

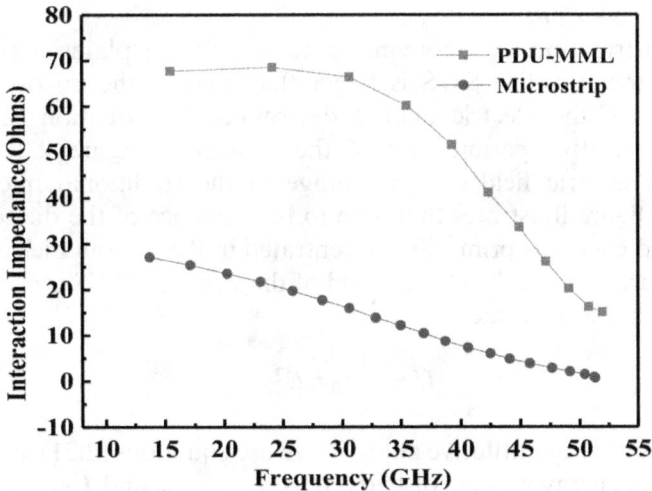

Figure 2.26. Comparison of the coupling impedances of the two structures. Reprinted from [9], with the permission of AIP Publishing.

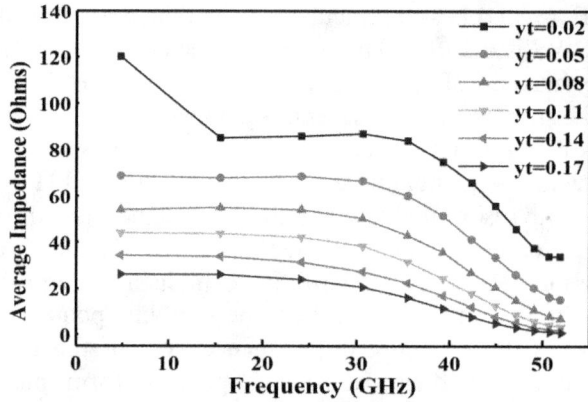

Figure 2.27. Coupling impedance curves of the strip line for different values of yt. Reprinted from [9], with the permission of AIP Publishing.

(PDU-MML) and microstrip meander line SWSs. In the Ka band (24–40 GHz), the coupling impedance of the strip-line type is approximately two to three times higher than that of the microstrip type. This indicates that eliminating the underlying dielectric substrate is indeed beneficial for increasing the coupling impedance, which is more conducive to the subsequent beam–wave interaction. To further study the attenuation of the coupling impedance of surface waves in the direction perpendicular to that of the transmission, the coupling impedance curve was simulated and plotted versus the yt value. As shown in figure 2.27, the coupling impedance reaches a maximum when yt is 0.02 mm. As yt gradually increases (the further the electron beam is from the surface of the strip line), the more the coupling impedance decreases. This provides a reference for the subsequent design of a simulation for beam–wave interaction.

2.2.1.4 Energy distribution

The energy distribution is a convincing reason that explains why the coupling impedance of the strip-line SWS is larger than that of the microstrip SWS. The energy storage of the electric field is determined by equation (2.20), where ε_r represents the relative permittivity of the dielectric. Figure 2.28(a) shows the distribution of electric field energy storage in the traditional microstrip planar structure. The figure illustrates that, due to the presence of the dielectric substrate, the electric field energy is primarily concentrated in the bottom dielectric region. As depicted in figure 2.28(b), the electric field of the strip-line SWS is evenly distributed on both sides of the strip line.

$$U = \frac{1}{2}\varepsilon_0\varepsilon_r E^2 \tag{2.20}$$

To perform a further quantitative analysis, we use equations (2.21) and (2.22), which approximate the energy storage of two SWSs. $U_{\text{microstrip}}$ and $U_{\text{PDU-MML}}$ denote the total energies stored by the microstrip and strip-line SWSs, respectively. In the simulation process of the eigenmode single-period structure, the total energy storage

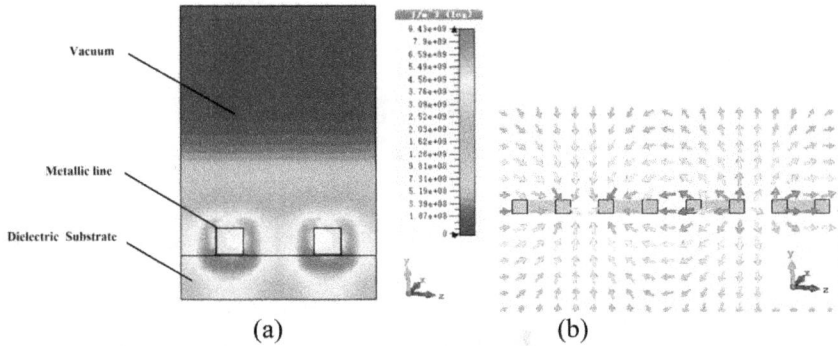

Figure 2.28. Electric field distribution diagrams. (a) Energy distribution of microstrip line SWS; (b) strip-line SWS. Reprinted from [9], with the permission of AIP Publishing.

calculation methods of the two structures are the same. However, due to the presence of the dielectric substrate (relative dielectric constant $\varepsilon_r = 4$) at the bottom of the microstrip SWS, E^2 is relatively small. The strip line is 2.5 times larger. The comparison of coupling impedances in figure 2.26 also reflects a similar trend:

$$U_{\text{microstrip}} \approx \frac{1}{2}\varepsilon_0\varepsilon_r E_{\text{down}}^2 + \frac{1}{2}\varepsilon_0 E_{\text{up}}^2 \approx \frac{1}{2}\varepsilon_0 E_{\text{up}}^2(1 + \varepsilon_r) \tag{2.21}$$

$$U_{\text{PDU}-\text{MML}} \approx \frac{1}{2}\varepsilon_0\left(E'_{\text{down}}\right)^2 + \frac{1}{2}\varepsilon_0\left(E'_{\text{up}}\right)^2 \approx \frac{1}{2}\varepsilon_0\left(E'_{\text{up}}\right)^2(1 + 1). \tag{2.22}$$

This result indicates that dielectric materials have the ability to absorb waves, meaning that EMWs tend to concentrate in the dielectric. In the planar microstrip SWS, the electric field is mainly concentrated in the dielectric substrate, resulting in a low coupling impedance in the beam–wave interaction region above the strip line. This is also one of the reasons for the use of the dielectric rods placed on both sides.

2.2.1.5 Transmission characteristics

The complete SWS designed in this chapter is shown in figure 2.29(a), including the periodic structure of 50 units, an attenuator, the strip-line tapered transition structure, and the tapered coaxial energy transmission windows. The selection of 50 periodic elements for the SWS is based on the subsequent simulation results of the beam–wave interaction. The use of too many or too few cycles is not conducive to signal amplification in TWTs. There is an optimal value for the number of elements.

The attenuator is an important part of the SWS. It is typically fabricated by steaming carbon on a dielectric rod (usually made from beryllium oxide, BeO) to impart a certain level of conductivity. As a result, it absorbs electromagnetic signals from all directions indiscriminately, exhibiting weak frequency selectivity. Although its main function is to attenuate the EMW, it has the following further functions: (1) suppressing the reflected wave and back waves to produce potential oscillation and maintain stable interaction; (2) extending the interaction length to further enhance the output power. Both of these functions are critical for a TWT.

(a)

(b)

(c)

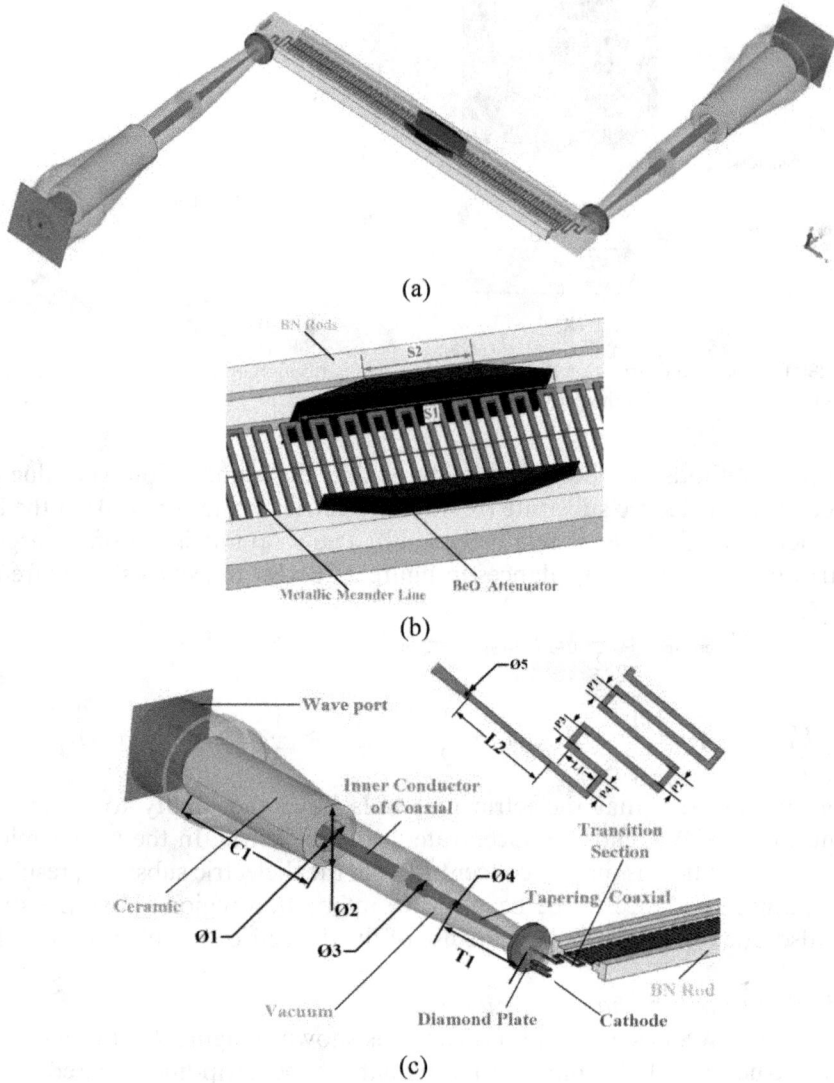

Figure 2.29. Transmission model of the strip-line SWS. (a) Overall structure, (b) attenuator; (c) tapered transitions and tapered coaxial windows. Reprinted from [9], with the permission of AIP Publishing.

The attenuators designed in this section are shown in figure 2.29(b), arranged in pairs on both sides of the strip line. The top view is an isosceles trapezoid. The lengths of the bottom and top are S_1 and S_2, respectively. The height is half that of the projecting part of the dielectric rod. The design of the attenuator's two sides aims to minimize the reflection of EMWs passing through the attenuator and achieve maximum attenuation.

Figure 2.29(c) shows the strip-line tapered transition structure and the tapered coaxial energy transmission window, which are placed in mirrored forms at the input and output ends of the SWS. Each period of the strip transitional structure presents

Table 2.3. The structural dimensional parameters of transition structure.

Structural parameter	Value (mm)	Structural parameter	Value (mm)
$C1$	7.5	$P1$	0.195
$T1$	4.6	$P2$	0.23
$\varnothing 1$	1.32	$P3$	0.21
$\varnothing 2$	2.56	$P4$	0.23
$\varnothing 3$	0.56	$L1$	0.52
$\varnothing 4$	0.42	$L2$	1.2

an unequal pattern of differing shapes, gradually becoming shorter and fatter. After the structural parameters are optimized, the input and output signal reflections can be significantly reduced. The purpose of the tapered coaxial energy transmission windows is to gradually transfer the signal from the thin strip line to the inner conductor of the Ka-band standard coaxial interface. The key structural parameters of this section are listed in table 2.3.

Following the establishment of the TWT model, the electromagnetic transmission characteristics of the planar strip-line SWS proposed in this chapter were studied using the CST Microwave Studio simulation software. In the field of TWTs, the research scope of electromagnetic transmission characteristics encompasses the following aspects: an EMW is fed in through the input port, transmitted along the slow-wave line, and fed out through the output port. In this process, echoes (reflections) caused by structural discontinuity are inevitable, as are insertion losses caused by metal loss and medium loss. These two characteristics are very important and are characterized by two parameters, S_{11} and S_{21}, respectively. Therefore, electromagnetic transmission characteristics are often referred to as S-parameter characteristics.

To examine the influence of the attenuator, this section studies the SWS of the TWT both with and without the attenuator. The materials for each part of the SWS are selected as follows: the strip line is molybdenum, the conductivity in the Ka frequency band is 2×10^7 S m^{-1}, and the dielectric rod is boron nitride. The attenuator is beryllium oxide coated with carbon. The dielectric constant is $\varepsilon_r = 6.5$, and the loss tan $\delta = 0.5$.

The final simulation S-parameter curve is shown in figure 2.30. The reflection S_{11} remains below -15 dB within the 34–40 GHz range, irrespective of the presence of the attenuator. This observation confirms the effective transition characteristics of the designed wedge-type attenuator, showing that it does not introduce additional reflection into the overall slow-wave circuit. Furthermore, the analysis indicates that the predominant cause of reflection can be attributed to the inherent discontinuity of the SWS itself. However, the magnitude of this reflection is within acceptable limits and meets the design requirements.

2.2.2 Coplanar waveguide SWS

Figure 2.31 shows a perspective view of the coplanar SWS presented in [10]. The structure consists of a dielectric substrate with a metal layer printed on top. The

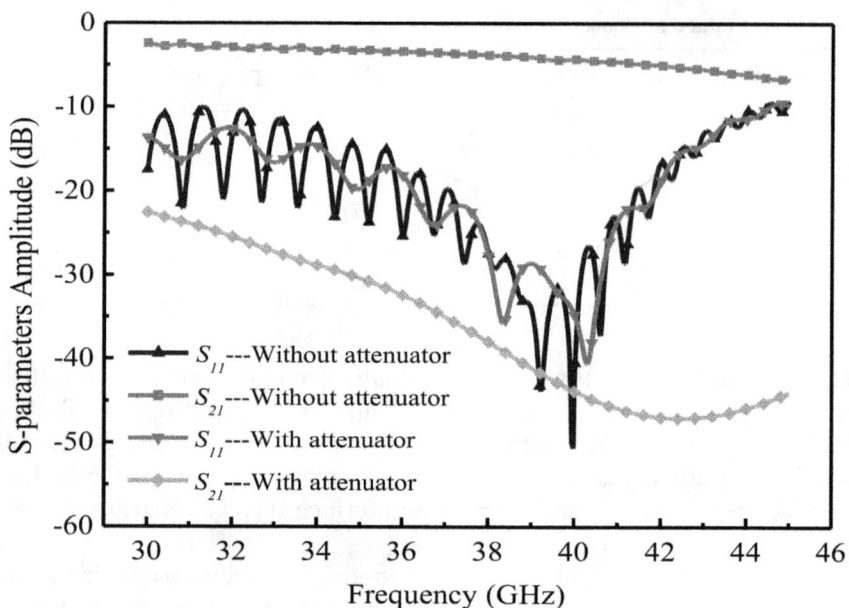

Figure 2.30. Transmission characteristic curves, divided into two cases, namely with and without attenuators. Reprinted from [9], with the permission of AIP Publishing.

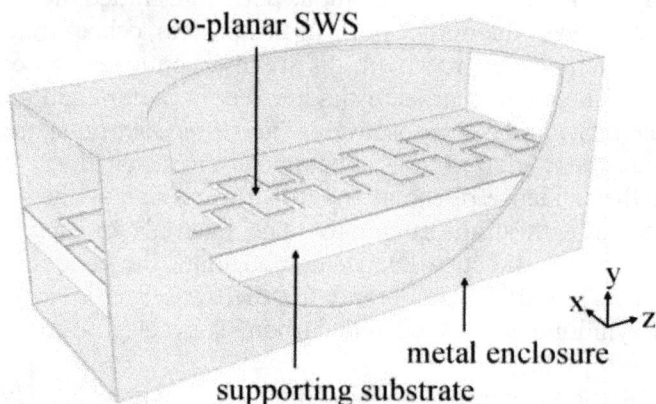

Figure 2.31. Perspective view of the coplanar SWS. © [2021] IEEE. Reprinted, with permission, from [10].

metal layer consists of three parts. The central part has fins on both sides, while the left and right parts are symmetrically aligned at the sides of the central pattern using a comb-like structure. Symmetric U-shaped grooves are formed between the central pattern and the side patterns. The whole SWS is located at the center of a metal enclosure.

It has been demonstrated that the structure possesses significantly higher coupling impedance and simpler input/output couplers compared to those of the microstrip meander-line SWS. In addition, the proposed structure exhibits a notable reduction in dielectric charging issues and improved heat dissipation, which is primarily attributed to the extensive coverage of metal on the top surface of the substrate and good metal contact with the metal enclosure. Furthermore, it is easier to fabricate as it eliminates the need for backside processing.

As outlined in reference [10], the proposed SWS has a complementary structure to that of the double MML-SWS. These two structures demonstrate nearly identical dispersion properties. Consequently, the dispersion characteristic of the proposed SWS can be predicted by analyzing the characteristics of the double MML-SWS.

2.2.3 Periodic slot SWS on a metal film

The periodic slot SWS comprises a metal film engraved with a periodic flat row of slots perpendicular to the electron beam transmission direction and a metal enclosure (figure 2.32(a)). The metal film is welded in parallel to the middle position of the metal shielding shell, just like the topology of the coplanar SWS presented in section 2.2.2. The sheet electron beam sweeps across the surface of the periodic flat row slot, stimulating and interacting with the surface wave, producing the EMW signal at the output end. A transmission model was built based on the use of 20 periods. The S-parameters were calculated, which showed low loss and wideband matching in the frequency range of 0.2–0.3 THz. The insertion loss was calculated to be as low as 0.16 dB mm^{-1} at 0.22 THz when the conductivity was set to 3×10^7 S m^{-1} (figure 2.32(b)).

The periodic slot SWS employs 2π-TM$_{n0}$ as the operating mode, and the operating frequencies of the different eigenmodes follow an approximate frequency multiplication relationship, as shown in figure 2.33. The voltage can be tuned to operate the THz radiation source at different frequency points, effectively enhancing the operating frequency without reducing the overall size. At increased voltage, the THz radiation source can function in higher harmonic modes, enabling the operating frequency to exceed 1 THz. This capability positions it as a promising miniaturized planar THz radiation source.

When operating as a backward wave oscillator, the structure can operate in the fundamental mode at an operating voltage of 3.7 kV and generate an output power of 0.225 W at 0.291 THz. Furthermore, the third harmonic output power can reach 3.48 W at 0.681 THz when the operating voltage is increased to 14.6 kV.

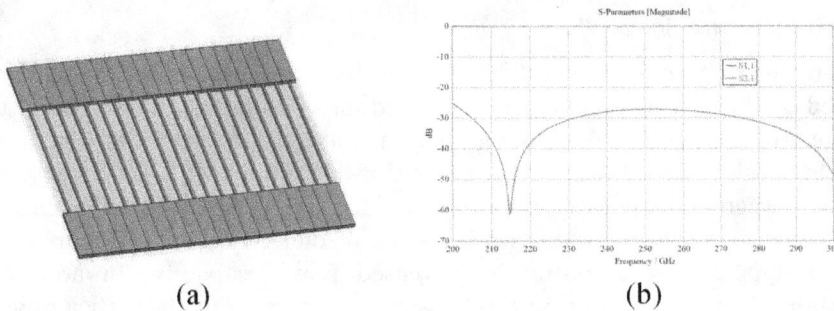

(a)　　　　　　　　　　　　　　　(b)

Figure 2.32. (a) Perspective view of the periodic slot SWS and (b) its transmission characteristics.

(a)

(b)

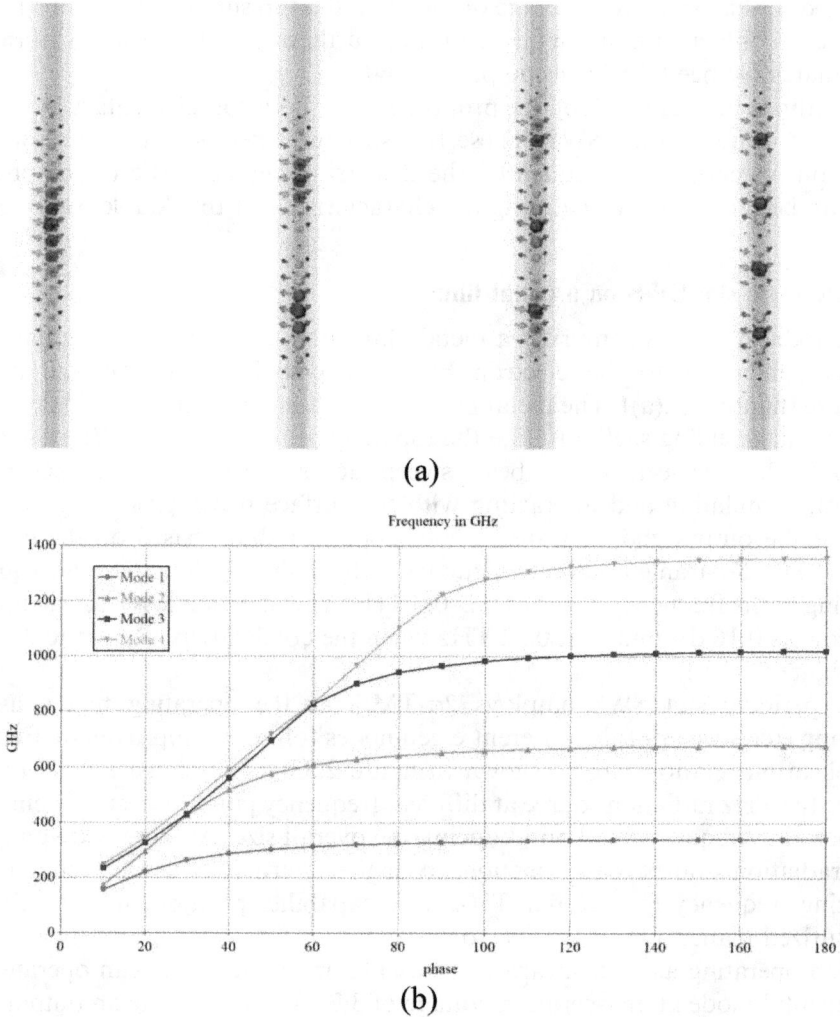

Figure 2.33. Operating modes of the periodic slot SWS: 2π-TM_{10}, 2π-TM_{20}, 2π-TM_{30}, 2π-TM_{40}. (a) E-field distribution; (b) dispersion curves.

2.2.4 Meander slot-line SWS

The all-metal SWS consists of a thin metal sheet on which a periodic meander slot is engraved by laser. A metal enclosure, divided into two halves, is fabricated using a computer numerical control (CNC) milling machine. An appropriate selection of the metal sheet thickness allows the meander slot-line SWS (MSL-SWS) to achieve a balance between performance and mass production efficiency [11]. The elimination of the dielectric substrate addresses the issue of dielectric charge accumulation in microstrip-type SWSs, resulting in increased power capacity. In addition, the application of copper sputtering after assembly serves to further efficiently reduce ohmic loss.

2.2.4.1 Model of the meander slot-line SWS

Figure 2.34(a) shows a single-period MSL-SWS model. The gaps between the enclosures and the metal sheet can be utilized as electron beam tunnels. Figures 2.34(b) and (c) are parametric models of the meander slot-line and the enclosure, respectively.

2.2.4.2 High-frequency characteristics of the meander slot-line SWS

Figure 2.35(a) depicts the ω–β diagram of the MSL-SWS. The diagram shows the first two eigenmodes and the 8.8 kV beam line, which intersects the forward wave region of mode 1 at around 40 GHz. As the beam line has no intersection point with mode 2 at the backward wave region, backward wave oscillation is unlikely to occur during forward wave amplification. Figure 2.35(b) depicts the curves of normalized phase velocity and coupling impedance. In the frequency range of 36–42 GHz, the dispersion curve is relatively flat, and the MSL-SWS has an on-axis coupling impedance of approximately 12–34 Ω at 0.1 mm from the slot-line surface.

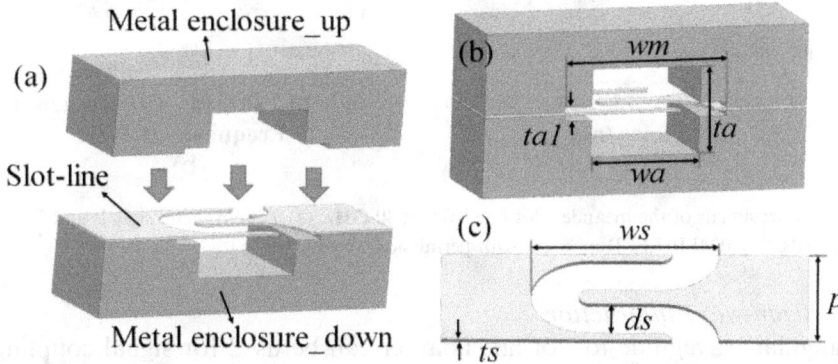

Figure 2.34. (a) Assembly of the three-dimensional model of a single-period meander slot-line SWS. (b) General and (c) detailed dimensional parameters of a single-period meander slot-line SWS. © [2024] IEEE. Reprinted, with permission, from [11].

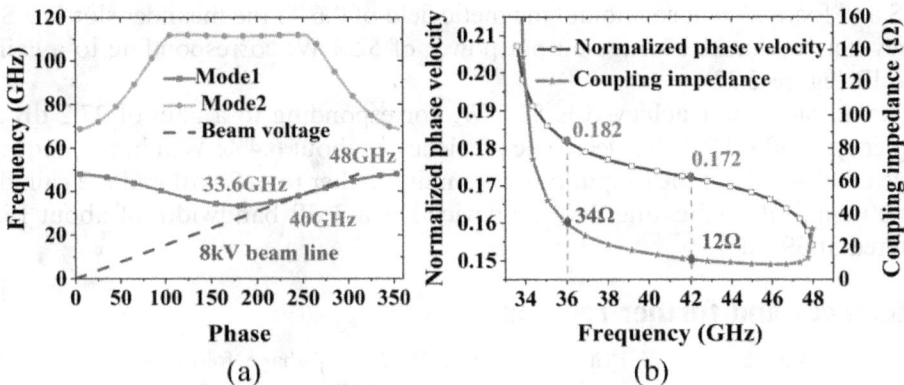

Figure 2.35. (a) Dispersion characteristics of the MSL-SWS. (b) Normalized phase velocity and coupling impedance of mode 1. © [2024] IEEE. Reprinted, with permission, from [11].

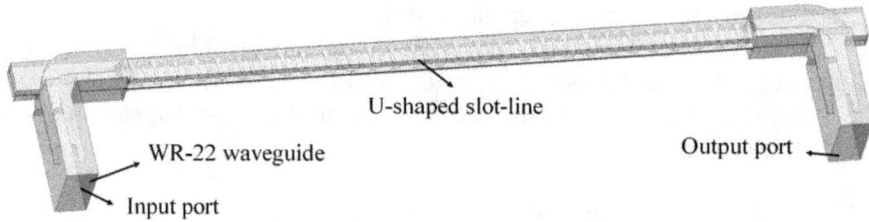

Figure 2.36. Diagram of the MSL-SWS model of beam–wave interaction. © [2024] IEEE. Reprinted, with permission, from [11].

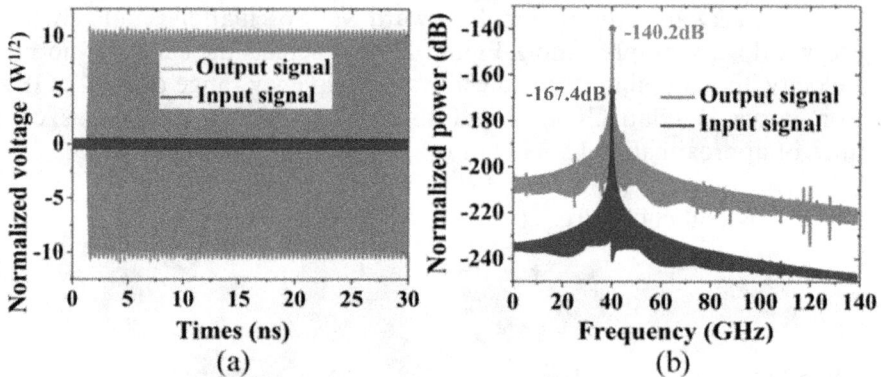

Figure 2.37. Port signals of the meander slot-line SWS at 40 GHz. (a) Time-domain signals and (b) frequency domain signals. © [2024] IEEE. Reprinted, with permission, from [11].

2.2.4.3 Beam-wave interaction

A rectangular waveguide-to-slot-line adapter can be used for signal coupling. The model for beam–wave interaction investigation shown in figure 2.36 comprises a pair of WR-22 standard rectangular waveguide to slot-line adapters and 29 uniform-period meander slot-line SWSs. Driven by an 8.2 kV, 0.1 A sheet beam with a cross-sectional area of 0.8 mm × 0.2 mm, which is positioned at 0.2 mm above the SWS and focused by a solenoidal magnetic field of 0.6 T, the meander slot-line SWS achieves a maximum average output power of 52.8 W, corresponding to a gain of 26.4 dB (figure 2.37).

The output power achieved is 52.5 W, corresponding to a gain of 27.2 dB at a frequency of 40 GHz. The electronic efficiency is about 6.4%. Within the frequency range of 36–42 GHz, the output power remains higher than 26 dB and the gain drop is less than 3 dB, indicating that the device has a 3 dB bandwidth of about 15.4% centered at 39 GHz.

References and further reading

[1] Liu Y, Yue L, Tian Y, Xu J and Wang W 2012 V-shape folded rectangular groove waveguide for millimeter-wave traveling-wave tube *IEEE Trans. Plasma Sci.* **40** 1027–31
[2] Lai J *et al* 2012 W-Band 1-kW staggered double-vane traveling-wave tube *IEEE Trans. Electron Devices* **59** 496–503

[3] Xu X *et al* 2011 Sine waveguide for 0.22-THz traveling-wave tube *IEEE Electron Device Lett.* **32** 1152–4

[4] Gottfried A H, Jasper L J and Tancredi J J 1976 Planar ring bar traveling wave tube[P] *US Patent* 3971966

[5] Wessel-Berg T 2006 Basics of radial sheet beam interactions with potential applications in the microwave K- and W-bands *AIP Conf. Proc.* **807** 55–64

[6] Wang S *et al* 2013 Study of a log-periodic slow wave structure for Ka-band radial sheet beam traveling wave tube *IEEE Trans. Plasma Sci.* **41** 2277–82

[7] Wang S *et al* 2013 A novel angular log-periodic micro-strip meander-line slow wave structure for low-voltage and wideband traveling wave tube *2013 IEEE 14th Int. Vacuum Electronics Conf. (IVEC) (Paris, France)* pp 1–2

[8] Xu D *et al* 2020 Theory and experiment of high-gain modified angular log-periodic folded waveguide slow wave structure *IEEE Electron Device Lett.* **41** 1237–40

[9] Wang H, Wang Z, Li X, He T, Xu D, Gong H, Tang T, Duan Z, Wei Y and Gong Y 2018 Study of a miniaturized dual-beam TWT with planar dielectric-rods-support uniform metallic meander line *Phys. Plasmas* **25** 063113

[10] Zhao C, Aditya S and Wang S 2021 A novel coplanar slow-wave structure for millimeter-wave BWO applications *IEEE Trans. Electron Devices* **68** 1924–9

[11] Wang Y *et al* 2024 Investigation of a novel planar meander slot-line slow-wave structure *IEEE Electron Device Lett.* **45** 476–9

Chapter 3

Planar electron optical systems

Compared with cylindrical electron beams, sheet electron beams feature larger cross-sectional areas, allowing for larger currents that have the same space-charge force. Therefore, sheet electron beams can be effectively utilized in high-power and high-frequency vacuum electron devices. Since the late 1990s, sheet electron beams have been a focal point of research, particularly in the realm of high-power microwave and millimeter-wave devices. Systems designed to generate and sustain the stable transmission of a sheet electron beam are commonly referred to as planar electron optical systems (EOSs).

3.1 Introduction

Various methods exist for generating a sheet electron beam, including the use of an anode slit, the magnetic quadrupole extension circular electron beam, and the elliptic helix tube compression circular electron beam. However, the most convenient method involves employing a convergent cylindrical cathode for the direct emission of the sheet electron beam.

The sheet beam electron gun consists of a section of a convergent coaxial cylindrical diode positioned outside the cathode and inside the anode. The shapes of the focusing electrode (also known as the beam-forming electrode (BFE)) and anode can be calculated using the Pierce gun theory. A sheet beam electron gun has two pairs of boundaries, i.e. one pair in each of the horizontal and vertical directions, necessitating the use of two pairs of electrodes for electron beam focusing, as depicted in figure 3.1(a). If two pairs of focusing electrodes share the same potential, they can be combined into a single electrode. Figure 3.1(b) illustrates an electron beam gun that has a single focusing electrode.

The presence of two pairs of boundaries makes the stable transmission of a sheet electron beam at high current more difficult than the transmission of a cylindrical electron beam, resulting in the well-known diocotron instability, especially when the sheet electron beam is transmitted in a solenoidal magnetic field. As research has

doi:10.1088/978-0-7503-5452-3ch3

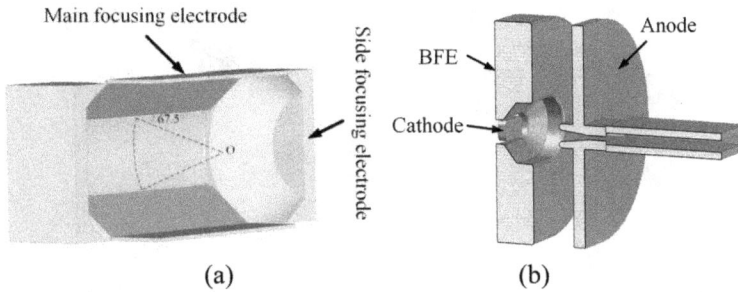

Figure 3.1. Sheet beam electron gun with (a) two pairs of beam-forming electrodes [1] and (b) a single beam-forming electrode [2].

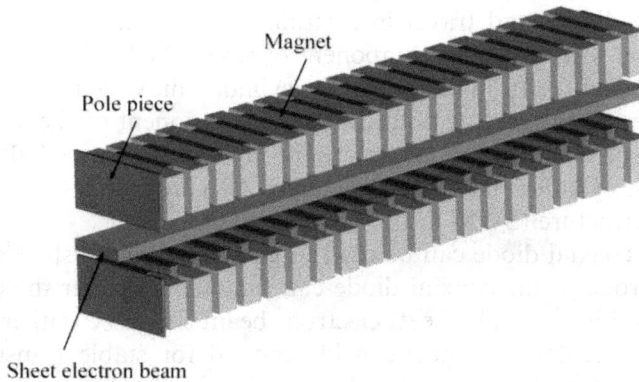

Figure 3.2. Periodic cusped magnetic (PCM) field focusing system for a sheet electron beam.

progressed, various focusing methods have been proposed, including the super magnetic field method, the conductor boundary method, and the ion neutralization method. However, it is worth noting that the effectiveness of these methods is not perfect. Today, the magnetic field focusing method is the most extensively employed; it includes various techniques such as periodic cusped magnetic (PCM) field focusing [3] (figure 3.2), periodically cusped-magnetic -periodic-quadrupole-magnetic (PCM-PQM) mixed focusing [4], wiggler magnetic field focusing [5], and closed PCM focusing [6]. While these magnetic structures are lightweight and compact, enabling the stable transmission of banded electron beams over long distances, they are still in the early stages of development and are yet to reach practical application.

Multistage depressed collectors for sheet electron beams help to improve overall device efficiency. Figure 3.3 illustrates a four-stage depressed collector that employs an elliptic cylindrical cavity as the electrode. Each electrode inlet is designed in the shape of an ellipse.

3.2 Radial planar EOSs

Utilizing the same cylinder as a reference, the distinction between a cylindrical electron beam and a radial beam lies in their respective emission characteristics. In

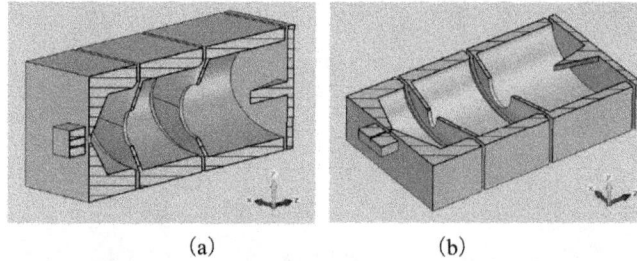

(a)   (b)

Figure 3.3. Sheet beam multistage depressed collector. (a) Longitudinal profile of the collector in the y–z-plane; (b) longitudinal profile of the collector in the x–z plane.

the case of the cylindrical electron beam, electrons are emitted from the bottom surface of the cylinder and travel in a straight line in the axial direction. These electrons interact with the axial component of slow EMWs. In contrast, the radial electron beam is emitted from the side of the cylinder, moving in a straight line in the radial direction and interacting with the radial component of the slow EMW. The interaction between the two can take place throughout the entire 360° perpendicular to the axis of the cylinder. This extensive interaction area allows the radial length of the slow-wave structure (SWS) to be shorter.

Generally, a coaxial diode can be used to generate a radial sheet electron beam. The inner electrode of the coaxial diode can function as either the cathode or the anode; the resulting radial sheet electron beam may be either divergent or convergent. The Brillouin magnetic field required for stable transmission of the radial banded electron beam can also be calculated. The designs for both radially divergent electron beam and radially convergent electron beam EOSs are elaborated on below.

3.2.1 Radial divergent sheet beam EOSs

A radially divergent electron beam refers to an electron beam emitted from the cathode at the center of the system. It is then accelerated by the anode, moving in a straight line in a radial direction. The thickness of the electron beam remains constant during its motion, but the current density of the electron beam gradually decreases as its radius increases. The radial divergent sheet beam EOS is a concentric cylindrical system that consists of an electron gun, a focusing system, and a collector (from the inside out) [7].

3.2.1.1 Electron gun

A short coaxial diode with an inner cathode is used as the model of the radially divergent sheet beam electron gun. It has been demonstrated that the maximum current transferable from the hot cathode to the anode with positive potential in a high vacuum environment is constrained by the space charge of the electron, a limitation known as the space-charge-limited current. In this case, increasing the operating temperature of the cathode does not result in an increase in the cathode current. For coaxial diodes of infinite length, the space-charge-limited current that can be emitted is determined by Langmuir–Blodgett's law [7]:

$$i = \frac{2\sqrt{2}}{9}\sqrt{\frac{e}{m}}\frac{V^{3/2}}{(r\beta^2)} \tag{3.1}$$

where i is the current per unit length in the axial direction; V is the anode voltage; e and m are the charge and mass of the electron, respectively; β is a structural correction factor; and r is the cathode radius. A comprehensive analysis of the calculation has been presented in the literature, and the details are not reiterated here. For a given voltage V and current i, the relationship between r and β can be derived using equation (3.1). Thus, the size of the electron gun can be roughly determined.

However, it is crucial to note that the application premise of equation (3.1) assumes that the length of the coaxial diode is infinite. Therefore, this equation can only serve as a preliminary reference for the actual design of the electron gun. For coaxial diodes with finite length, the density of the space-charge-limited current can be approximated by [8]:

$$J_c = \left(1 + \frac{2R_c^2}{W^2}\right)\frac{4\varepsilon_0}{9}\sqrt{\frac{2e}{m}}\frac{V^{3/2}}{D^{1/2}R_c^{3/2}}\left(\frac{1}{\ln(R_a/R_c)}\right) \tag{3.2}$$

where W is the length of the diode; D is the distance between the anode and the cathode; R_a and R_c are the radius of the anode and that of the cathode, respectively; ε_0 is the vacuum intermediate electrical constant; and V is the anode voltage. Formula (3.2) is valid under the condition that $W \gg D$, and it is applicable in the design of conventional sheet-beam electron guns that have a large aspect ratio.

If the infinitely long coaxial diode is regarded as a one-dimensional structure, the coaxial diode of finite length is considered to be a 2D structure, and the short coaxial diode used in the radial sheet beam electron gun represents a 3D structure, it is fundamentally challenging to solve the space-charge-limited current of such a structure. Fortunately, the particle simulation method can be employed for precise design based on the aforementioned theories. The common design software used for electron guns includes EGUN, MAGIC, CST STUDIO SUITE, etc. These software packages can conveniently calculate the trajectories of charged particles in the electric field and the magnetic field. The process of particle emission is solved by gun iteration. The solution process is as follows: the solver first calculates the initial potential distribution in the electron gun region. Subsequently, it records the trajectories and space charge distributions of the emitted particles. It then uses the space charge to correct the charge vector and calculate the new electric field potential until the results converge.

Figure 3.4(a) depicts a cylindrical coaxial diode model where the inner cylinder serves as the cathode, the side functions as the transmitting surface, and the outer ring acts as the anode. The height of the diode is 2 mm, and the distance between the anode and the cathode is 2 mm. Employing space charge to limit emission, figure 3.4(b)

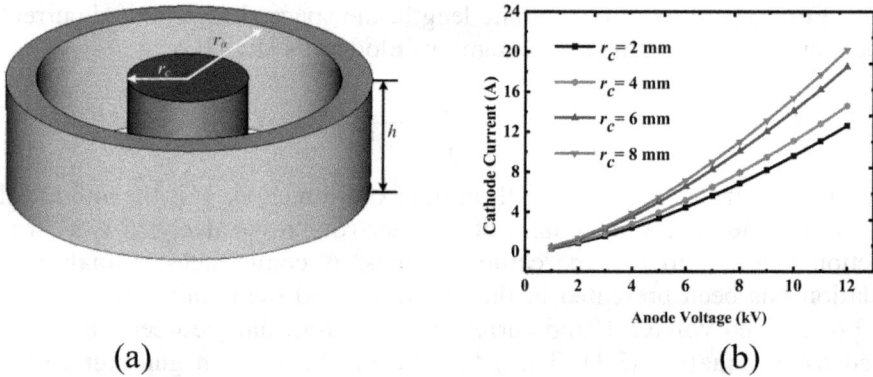

Figure 3.4. (a) Short coaxial cylindrical diode model with an inner cathode and an outer anode; (b) its characteristic voltammetry curves.

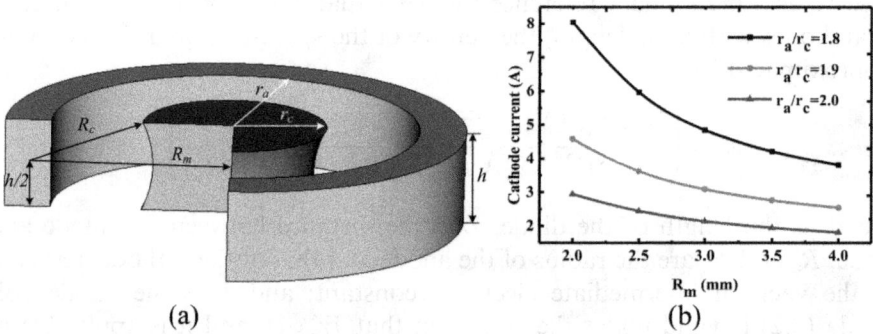

Figure 3.5. (a) Coaxial diode model with an arc-shaped cathode emission surface; (b) its characteristic voltammetry curve.

illustrates the emission current curves obtained using different cathode radii and anode voltages. It is evident that when the anode voltage is less than 2 kV, changing the cathode radius impacts the emission current by less than 2 A. When the voltage is less than 1 kV, the current remains below 1 A.

The cathode current of the diode can be effectively enhanced by increasing the area of the emitting surface or by reducing the distance between the anode and cathode while keeping the anode voltage constant. Building upon this concept, a cylindrical coaxial diode with an arc-cathode emission surface can be employed. An additional advantage of using a curved emitting surface is that the electron beam converges in the thickness direction after leaving the cathode, facilitating focusing.

A coaxial cylindrical diode with a curved cathode emission surface was built to quantitatively study the influence of the radius of the cathode emission surface on the cathode emission current (figure 3.5(a)). The magnitude of the cathode current was obtained by computer simulation for different values of R_m and the anode–cathode distance, as shown in figure 3.5(b). In the calculation, the anode voltage was set to 1600 V. The height of the diode matched that of the diode shown in figure 3.4(a). It was observed that for small R_m values, corresponding to large radii of the emitting surface, the cathode current of this diode surpassed 3 A with ease. This significantly exceeds the cathode current of conventional cylindrical coaxial diodes. In addition, with an increase in R_m, the cathode current decreased rapidly and then gradually, corresponding with the qualitative analysis above. It can be asserted that, compared to the ordinary cylindrical coaxial diode, the curved-cathode diode significantly enhances the cathode emission current while remaining within the same dimensions.

A radially divergent sheet beam electron gun including the cathode, focusing electrode, and anode, designed using a coaxial arc-cathode diode, is shown in figure 3.6. The blue component at the center of the figure is the cathode, the gray components above and below the cathode are the beam-forming electrodes, and the yellow components at the outer sides are the annular anodes. This figure also provides the dimensions of each component. It is now necessary to study the influence of various parameters on the cathode current in order to design the radial divergent banded beam electron gun with the best structure and performance.

Variations in the cathodic current corresponding to different structural dimensions are depicted in figure 3.7, as discussed in the following. From figure 3.7(a), it is clear that the cathode emission current increases with the augmentation of w_g. This phenomenon can be attributed to the intensified electric field at the cathode surface resulting from the closer proximity of the anode tip to the cathode emission surface. However, due to the existence of the tip effect, the electric field at the anode tip is stronger than that at other positions. This disparity leads to local bending of the equipotential surface on the cathode, causing electrons at the anode tip to acquire a higher transverse velocity compared to those at other positions. Consequently, this imbalance results in electron chaos. Figure 3.7(b) shows that as l is incremented from 0.8 mm to 1.2 mm, the current exhibits a predominantly monotonic decrease. However, the change in the current value is minimal, which can be attributed to the diminishing electric field on the cathode surface. In figure 3.7(c), cathodes of greater height possess a larger emission area, consequently allowing for the emission of a higher current. Nonetheless, as the cathode height increases, the electric field

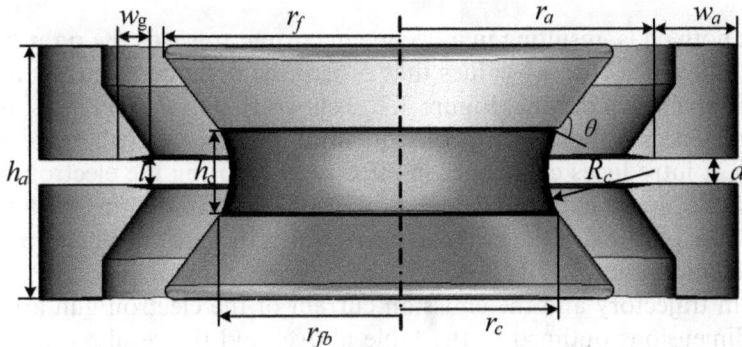

Figure 3.6. Profile of radial divergent electron gun structure. © [2012] IEEE. Reprinted, with permission, from [9].

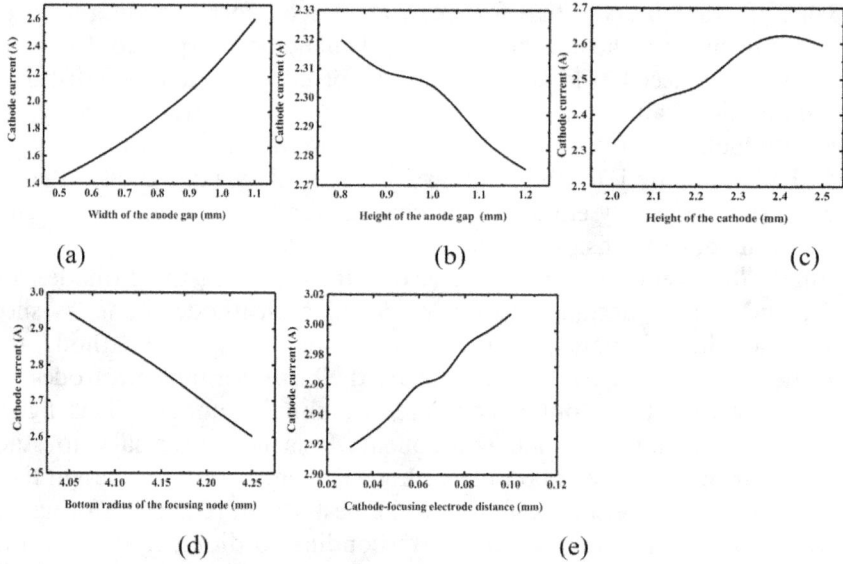

Figure 3.7. Graphs of the cathode current versus (a) w_g, (b) l, (c) h_c, (d) r_f, and (e) d_{cf}. © [2012] IEEE. Reprinted, with permission, from [9].

Table 3.1. Parametric values of the divergent radial beam electron gun.

Parameter	Value	Parameter	Value
R_c	1.8 mm	w_g	0.95 mm
r_c	4 mm	d_{cf}	0.05 mm
h_c	2 mm	h_a	6.1 mm
r_f	5.66 mm	d	0.6 mm
r_a	6.8 mm	l	0.8 mm
r_{fb}	4.05 mm	θ	67.5°
w_a	1.5 mm		

weakens at both ends, resulting in a decline in current towards the right-hand end of the curve. In figure 3.7(d), r_{fb} values that exceed the cathode radius by a significant margin exhibit a sharp decline. Figure 3.7(e) shows that as d_{cf} increases, the cathode current experiences a slight, yet inconsequential, rise. In fact, when the d_{cf}/h_c value exceeds 0.1, it introduces challenges in effectively focusing the electron beam.

Based on the aforementioned calculation results and considering both the cathode current and the position of the electron beam waist, the parameters were determined and are presented in table 3.1.

The beam trajectory and the emission current of the electron gun are calculated using the dimensions outlined in the table above, and the results are illustrated in figure 3.8. The depiction reveals a uniform distribution of electrons in the radial direction and a gradual decrease in beam thickness in the vertical direction. The

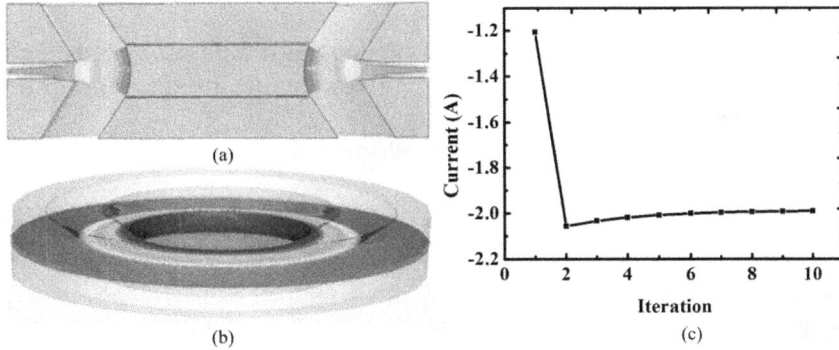

Figure 3.8. Electron gun calculation results from blue to red, the beam voltage increases gradually. (a) Side view; (b) perspective view; (c) cathode emission current curve. © [2012] IEEE. Reprinted, with permission, from [9].

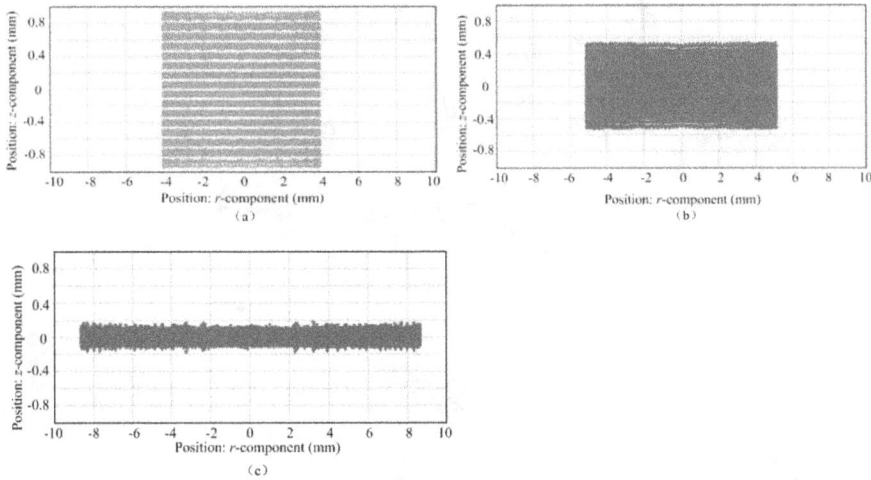

Figure 3.9. Cross-sections of electrons at different radial positions. (a) Cathode surface; (b) $r = 5.2$ mm; and (c) beam waist. © [2012] IEEE. Reprinted, with permission, from [9].

emission current curve of the electron gun stabilizes at approximately 1.99 A after 10 iterations. At this point, the electron current density at the cathode is 4 A cm^{-2}, and the perveance of the electron gun stands at 3.125×10^{-5} P.

The vertical compression process that occurs during the movement of the electron beam from the cathode surface to the beam waist is illustrated in figure 3.9. Notably, the electron beam exhibits a smooth laminar flow throughout its trajectory. At the beam waist, the maximum thickness of the electron beam is observed to be less than 0.4 mm, which corresponds to the specified design requirements.

3.2.1.2 The use of a magnetic field for focusing

Typically, in the longitudinal structure of a traveling-wave tube (TWT), a magnetic field aligned in either the same or the opposite direction to that of the electron beam is employed for focusing. In accordance with this concept, focusing a radial electron beam necessitates only the application of a vertical magnetic field force. Therefore, a radial magnetic field can serve as the focusing magnetic field. Analyzing the motion of electrons in the radial magnetic field can offer valuable insights into the design of the magnetic field [7].

When a charged particle moves in an electromagnetic field, it experiences the Lorentz force, defined as:

$$m\vec{v} = -e(\vec{E} + \vec{v} \times \vec{B}) \tag{3.3}$$

In the case of the axisymmetric electron beam depicted in figure 3.10, where the space-charge field has only the z component E_z, the equation of motion for an electron in the cylindrical coordinate system can be expressed as:

$$\begin{cases} m\dfrac{d^2z}{dt^2} = -eE_z + er\dfrac{d\theta}{dt}B \\[2mm] m\left(\dfrac{d^2r}{dt^2} - r\left(\dfrac{d\theta}{dt}\right)^2\right) = 0 \\[2mm] m\left(2\dfrac{dr}{dt}\dfrac{d\theta}{dt} + r\dfrac{d^2\theta}{dt^2}\right) = -e\dfrac{dz}{dt}B \\[2mm] \dfrac{dr}{dt} = \sqrt{2\eta V_a} \end{cases} \tag{3.4}$$

where B is the radial component of the magnetic field and V_a is the anode voltage.

In accordance with Gauss's theorem, we get:

$$E_z = \pm\frac{I_0}{2\varepsilon_0(dr/dt)} \tag{3.5}$$

where I_0 is the current per unit angle.

By combining Busch's theorem with the third equation of (3.4), we can get:

$$\frac{d\theta}{dt} = \frac{\eta}{2\pi r^2}(\psi - \psi_0) \tag{3.6}$$

where ψ is the magnetic flux through the cross-section and ψ_0 is the magnetic flux at the initial position of the electron beam.

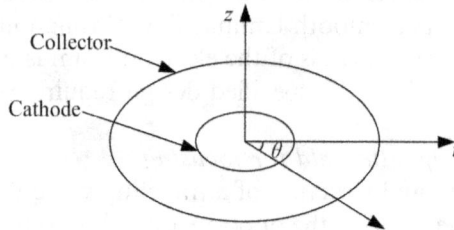

Figure 3.10. Radial electron beam in the cylindrical coordinate system. © [2012] IEEE. Reprinted, with permission, from [9].

By substituting (3.5) and (3.6) into the first equation of (3.4), the equation of motion for the electron beam along the z-axis can be obtained:

$$\frac{d^2z}{d^2t} - \eta^2 B^2\left(z - \frac{B_0 r_0}{Br}z_0\right) - \frac{\eta I_0}{2\varepsilon_0(dr/dt)} = 0. \tag{3.7}$$

The boundary trajectory equation of the electron beam in the radial direction can also be obtained:

$$\frac{d^2z}{d^2r} - \frac{\eta^2 B^2}{(dr/dt)^2}\left(z - \frac{B_0 r_0}{Br}z_0\right) - \frac{\eta I_0}{2\varepsilon_0(dr/dt)^3} = 0. \tag{3.8}$$

Let $d^2z/dr^2 = 0$. The equilibrium coordinates for the electron beam boundary trajectories can now be obtained as follows:

$$z_e = a_0 + \frac{B_0 r_0}{Br}z_0 \tag{3.9}$$

where $a_0 = I_0/2\varepsilon_0\eta B^2(dr/dt)$ is half the beam thickness of the electron beam.

Using the above formula, the expression for the Brillouin magnetic field can now be obtained:

$$B_b^2 = \frac{I_0}{2\varepsilon_0\eta a_0(dr/dt)}. \tag{3.10}$$

It is important to note that the result obtained from equation (3.10) is expressed in units of Gauss. The Brillouin focusing magnetic field is observed to be directly proportional to the angular current density and inversely proportional to the electron beam height and anode voltage. Through calculation, it can be determined that the Brillouin magnetic field required to focus a radial electron beam with a current of 2 A, a voltage of 1600 V, and a thickness of 0.4 mm is $B_b = 693$ Gauss.

Another technique for generating a radial magnetic field involves the use of a disc-shaped energized coil. When the coil approximates a circular shape, the magnetic fields on its surface can be treated as radial, uniform magnetic fields. Given that the radial magnetic field of a single disc coil rapidly diminishes with increasing distance from the coil surface, practical applications often utilize a pair of disc coils with opposing currents to achieve a larger radial magnetic field. Figure 3.11 is a schematic representation of the magnetic field generated by a pair of disc-shaped coils with opposing currents, as obtained through computer simulation. It can be observed that in the middle of the coil, the magnetic field is uniformly distributed in the radial direction and has almost no vertical component. In addition, B_r is proportional to the coil current and is constant in the middle range. The magnetic field generated by the coil in the vertical direction is much smaller than that in the horizontal direction; it is approximately one percent of the latter. The angular magnetic field component is merely one thousandth of the horizontal component. The main parameters of the coil in the figure are as follows: the inner

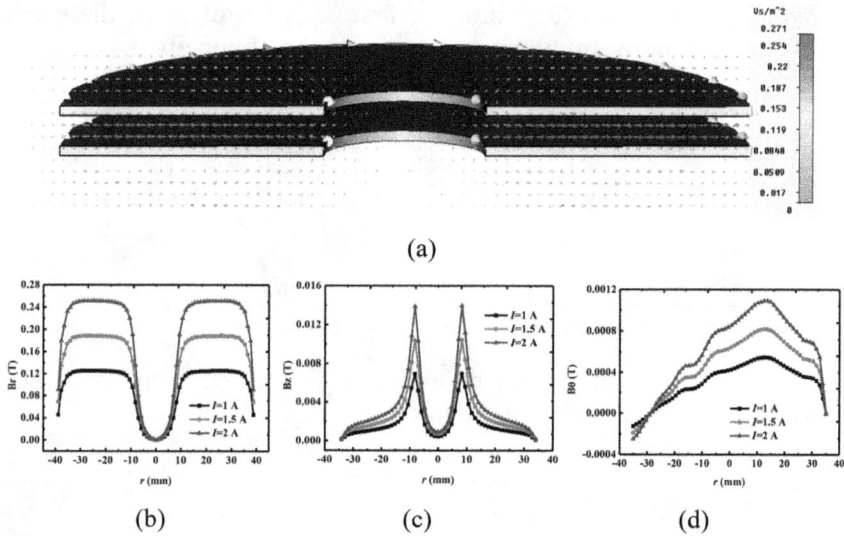

(a)

(b)   (c)   (d)

Figure 3.11. Magnetic field produced by a pair of plane coils. (a) Vector pattern; (b) radial component; (c) axial component; (d) angular component. © [2012] IEEE. Reprinted, with permission, from [9].

diameter of the coil is 8.2 mm, the outer diameter of the coil is 35 mm, the coil current is 2 A, and the distance between wire packets is 2 mm.

Figure 3.11(c) shows the vertical component of the magnetic field in the central plane of the two coils. It is evident that the flow of the radial electron beam between two metal plates can be investigated using the electron gun and radial focusing magnetic field designed above. The starting position of the magnetic field needs to be carefully adjusted, as any discrepancy between the magnetic field alignment and the electron beam may result in pulsation of the electron beam. Given a tunnel height of 0.6 mm for the radial electron beam, it navigates the drift tunnel seamlessly under a radial magnetic field exceeding 650 Gauss, as determined by the previously calculated Brillouin focusing magnetic field. Throughout this process, the thickness of the electron beam remains below 0.4 mm. Notably, the radial electron beam's advantageous characteristic lies in its singular pair of vertical boundaries, rendering it more stable and easier to focus. The flow of the electron beam in the drift channel is shown in figure 3.12. Figure 3.12(a) shows a side view of the electron beam trajectory, where the focusing magnetic field used is 1.23 times greater than the Brillouin field. The electron beam is well constrained in the vertical direction; the duty cycle is approximately 0.55 and the flow rate is 100%. Figure 3.12(b) shows the top view of the electron beam trajectory. The electrons travel along a linear path in the radial direction, smoothly traversing the entire drift channel and demonstrating remarkable stability.

3.2.2 Radial convergent planar EOSs

The radial convergent electron beam is a radial electron beam that has a constant thickness, a decreasing front-face radius, and an increasing current density

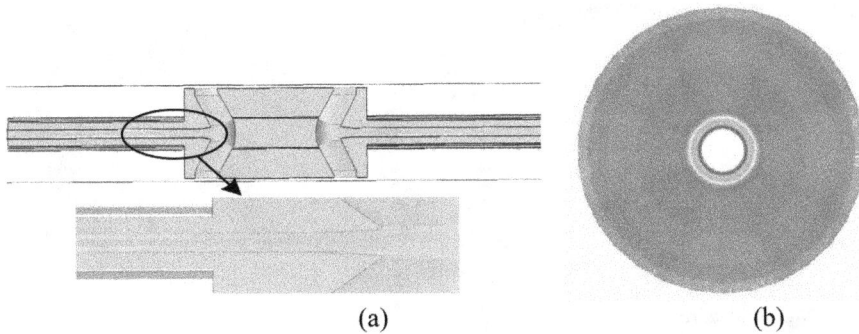

Figure 3.12. Trajectory diagram of a radial electron beam subjected to radial magnetic field focusing. (a) Side view; (b) top view. © [2012] IEEE. Reprinted, with permission, from [9].

throughout its transmission. Compared with the radially divergent electron beam, this electron beam exhibits a higher current density at the output end of the SWS, potentially leading to a stronger beam-to-wave interaction. However, it requires a stronger focusing magnetic field. The structure of the radially convergent electron beam closely resembles that of the divergent electron beam; the key distinction between them is that the radially convergent electron beam is transmitted inward from the outside. Specifically, the electron gun is situated at the outermost part of the structure, while the collector pole is positioned at the center [10].

This EOS is well-suited for compact radial beam TWTs characterized by their small size, especially in cases where the electron gun, including the cathode, cannot reach the necessary current levels. Furthermore, due to the passive characteristics of the magnetic field, it is difficult to obtain a uniform radial magnetic field at the center of radial emission electron beam TWTs that have a small radius, which impacts the formation and focusing of the electron beam. In such scenarios, the radial convergent electron beam outside the cathode can be adopted.

The radial convergent electron gun can be designed using a coaxial diode that has the cathode outside the anode. This design offers the advantage of a smaller size compared to that of the radial divergent electron gun while yielding a higher generated current. The primary goal of this section is to describe the implementation of a radial convergent electron beam for a miniature radial beam TWT. The key size parameters include a thickness of 0.04 mm, a current of 4 A, an initial radius of 20 mm, a transmission distance of 14 mm, and an electron beam voltage of 1600 V.

Given the electron beam's required thickness of merely 0.04 mm, cathode thicknesses in the range of 0.1 mm to 0.3 mm are considered. The corresponding electron beam compression ratio ranges from 2.5:1 to 7.5:1. The anode voltage is set to 1600 V, and the electron emission mode selected is space-charge-limited emission. It is evident that, compared with the divergent cylindrical coaxial diode described in the previous section, the convergent diode yields a larger emission current, even when the thickness is reduced to one-tenth of the thickness of the former. Figure 3.13 illustrates a radially convergent beam electron gun, including the cathode, focusing electrode, and anode; the corresponding dimensions are listed in table 3.2.

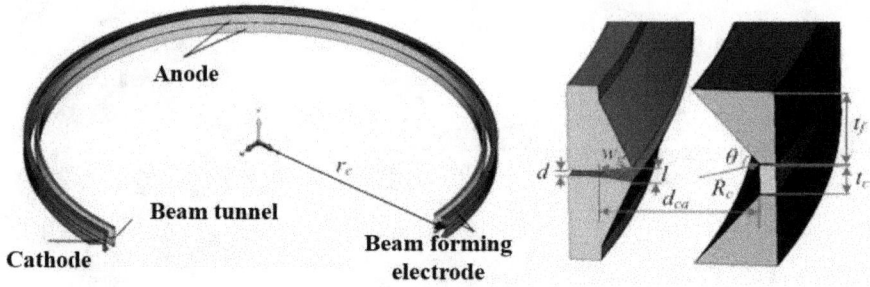

Figure 3.13. Model of a radial convergent electron gun model and its main dimensions. © [2014] IEEE. Reprinted, with permission, from [10].

Table 3.2. The parametric values of the radial convergent electron gun.

Parameter	Value	Parameter	Value
Rc	1 mm	w_g	0.2 mm
r_c	20 mm	d_{cf}	0.01 mm
t_c	0.2 mm	d	0.04 mm
t_f	0.5 mm	l	0.1 mm
d_{ac}	1.1 mm	θ	67.5°
w_a	1.5 mm		

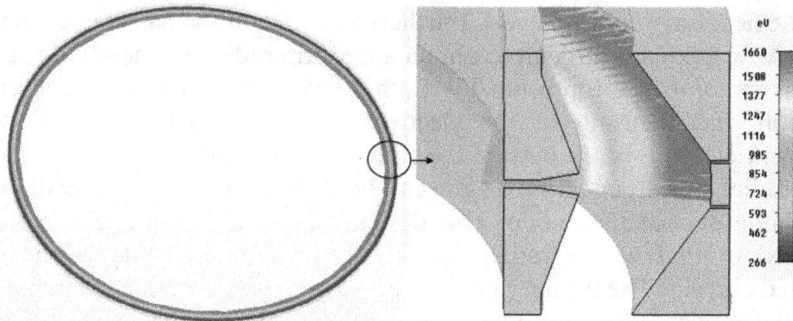

Figure 3.14. Calculated electron trajectory of a radial convergent electron gun. © [2014] IEEE. Reprinted, with permission, from [10].

Utilizing the dimensions specified in the above table and employing space-charge emission as the electron emission mode, the electron beam emission within the gun is depicted in figure 3.14. Notably, the emission of electrons throughout the entire gun region appears relatively uniform, and the vertical compression of the electron beam remains relatively stable. The current emitted by the cathode stabilizes at approximately 4 A. Consequently, calculations indicate that the emission current density and the perveance of the cathode amount to 15.9 A cm^{-2} and 6.25×10^{-5} P, respectively.

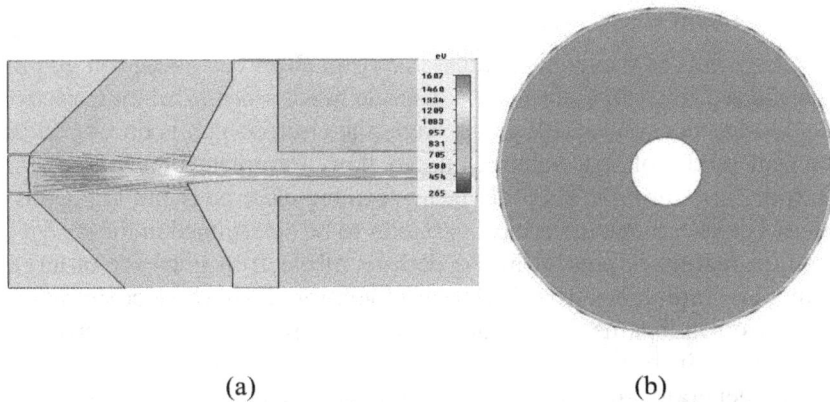

The radial convergent electron beam can be focused using the same radial uniform magnetic field as that used for the radially divergent electron beam. The required Brillouin magnetic field value can be obtained from the calculation results shown in section 3.2.1. It is important to note that as the current density of the radially convergent electron beam progressively increases during beam transmission, the required Brillouin focusing magnetic field also gradually intensifies from the outer to the inner regions. Therefore, in this context, I_0 should be the angular current density at the collector. By substituting the values of each parameter, the Brillouin focusing magnetic field strength of $B_b = 717$ Gauss can be calculated.

Using the electron gun and the radial focusing magnetic field, the flow of the radially convergent electron beam between two metal plates can be obtained for a tunnel height of 0.06 mm, which is aligned with the electron beam channel height of the SWS. The results indicate that the electron beam can pass through the drift channel smoothly when subjected to a radial magnetic field exceeding 0.06 T. Moreover, the beam thickness remains less than 0.04 mm throughout this process, as shown in figure 3.15.

3.3 Angular radial planar EOSs

The angular radial divergent or convergent electron gun is designed to produce an angular radial divergent or convergent sheet electron beam with the specific DC power and geometric size required for the angular log-period meander-line SWS. To further refine the emitted electron beam, a radially tunable PCM magnetic focusing system is employed. This system serves to control and maintain the shape of the angular radial divergent electron beam during its radial transmission process. The subsequent sections elaborate on research aspects pertaining to both the electron gun and the radially tunable PCM magnetic focusing system [11].

3.3.1 Angular radial electron gun

The key components of the angular radial divergent sheet beam electron gun include a ceramic–metal sealed shell, a hot wire, a cathode head, a beam-forming electrode, and an anode. During the theoretical design stage, the electron gun is developed based on the theoretical model of space-charge-limited flow. Parameters such as the shape of each electrode, including the cathode, beam-forming electrode, and anode, as well as the distances between them, are critical variables to be determined in the design process.

A common feature is present in the designs of electron guns for radial electron beams, angular radial electron beams, conventional sheet electron beams, and cylindrical electron beams. This commonality is reflected in the cross-sectional pattern of the electron gun. To elaborate, the same cross-sectional pattern can be employed to generate an electron gun for radial electron beams, angular radial electron beams, and conventional cylindrical electron beams by sweeping along different axes. The classic Pierce electron gun can be obtained by rotating the cross-sectional pattern (figure 3.16(a)) around the z-axis in the cylindrical coordinate system (figure 3.16(b)), while a sheet beam electron gun can be obtained by first mirroring about the x–y plane and then extruding the sum pattern (figure 3.16(c)) along the x-axis (as shown in figure 3.16(d)). The radial sheet beam electron gun can be obtained from the section shown in figure 3.16(c) by rotation around the R-axis (as shown in figures 3.16(e) and (f)). In the third scenario, if the rotational angle is less than 360°, an angular radial divergent electron gun model with a defined angle can be obtained.

Hence, the theoretical analysis of the radial sheet beam electron gun can draw insights from the analytical theory of axial electron guns. A well-established classical theory for such analysis is Pierce's design theory. For a comprehensive understanding of the theoretical derivation process, the reader is directed to the detailed information provided in [14].

In order to yield sufficient current, the angular radial electron gun necessitates either a high current emission density or the compression of electron beams. This process is similar to that of the Pierce electron gun; the key distinction between them lies in the formation of the 3D body through the sweep of the cross-sectional pattern (refer to figure 3.16 for an illustrative schematic diagram). In addition, the resulting electron beams are shaped and constrained in a manner specific to the angular radial electron gun design. Within the angular radial electron gun, the beam-forming electrode is oriented at a specific angle with respect to the axis of the central axial section, introducing an axial compression effect. The slope angle of the beam-forming electrode on the angled central section neutralizes the space-charge force exerted by the electron beam before and after reaching the beam position. This neutralization ensures the preservation of the electron beam's radial motion. Figure 3.17 illustrates a cathode with a cylindrical emission surface set at an angle of 8° on the cylinder.

Figure 3.18 shows the axial section and an angular section of the angular radial electron gun and their corresponding structural dimensions, respectively. For the specific structural dimensions, refer to table 3.3.

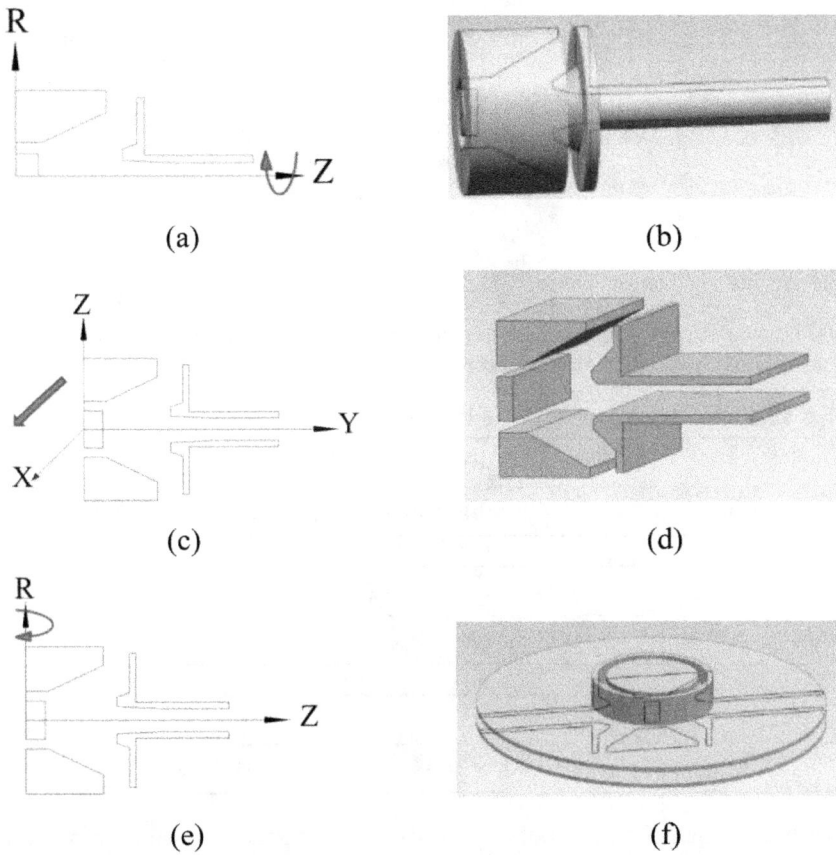

Figure 3.16. Schematic diagrams of different types of electron guns obtained using the same section and different rotational modes. (a) Cross-sectional pattern of a Pierce electron gun; (b) cylindrical electron gun obtained by rotating pattern (a) around the z-axis; (c) mirror pattern of (a) in Cartesian coordinates; (d) sheet electron beam gun obtained by extruding (c) along the x-axis; (e) mirror pattern of (a) in cylindrical coordinates; and (f) radial sheet electron beam gun obtained by rotating (e) around the r-axis.

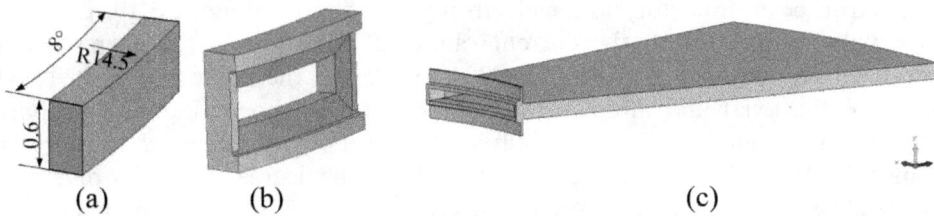

Figure 3.17. Structures of key components of the angular radial electron gun. (a) The cylindrical emission-surface cathode; (b) the beam-forming electrode; and (c) the anode and beam tunnel. © [2015] IEEE. Reprinted, with permission, from [11].

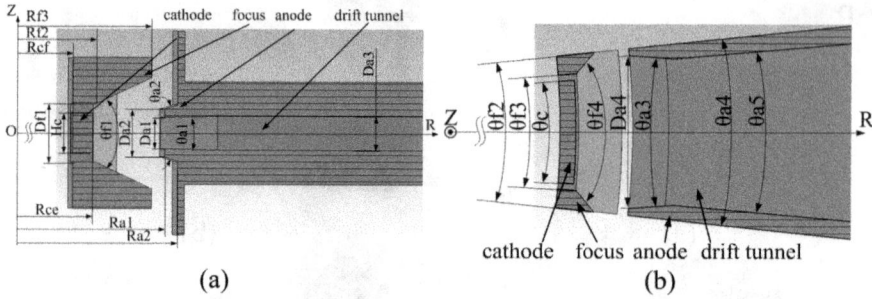

Figure 3.18. (a) Axial section and (b) angular section showing the structural dimensions of the angular radial electron gun. © [2017] IEEE. Reprinted, with permission, from [12].

Table 3.3. A list of the structural dimensions of the angular radial electron gun.

Hc	Df1	Rcf	Rf2	Rf3	Rce	Ra1
0.6 mm	0.9 mm	14.2 mm	14.55 mm	15.4 mm	14.5 mm	15.6 mm

Ra2	Da1	Da2	Da3	$\theta f1$	$\theta a1$	$\theta a2$
15.8 mm	0.42 mm	0.73 mm	0.52 mm	50.8°	3.1°	36.9°

θc	$\theta f2$	$\theta f3$	$\theta f4$	$\theta a3$	$\theta a4$	$\theta a5$	Da4
8°	12°	9°	99.8°	8°	12°	10°	3 mm

The emission density and perveance of the cathode emission surface can be calculated using equations (3.11) and (3.12):

$$J = \frac{I_0}{H_c \times \frac{\pi R_{ce}}{180} \times \theta_c} \tag{3.11}$$

$$P = \frac{I}{V^{\frac{3}{2}}} \times 10^6 (\mu P). \tag{3.12}$$

The electron beam trajectory in the electron gun is shown in figure 3.19. Under the anode voltage of 1700 V, the current obtained is 84.9 mA, the corresponding cathode current emission density is 6.99 A cm^{-2}, and the perveance is 1.11e-6 P. Images of the axial and angular electron trajectories show that the fan-shaped electron beam begins to diverge axially after traversing the channel for a certain distance (1~2 mm) (refer to figure 3.19(a)) and that the degree of angular divergence is relatively weak. A possible reason for this is that in the absence of a magnetic field, a large number of electrons are intercepted on the upper and lower (axial) walls of the channel in the process of movement. The remaining electron beam is less affected by the space-charge force in the angular direction.

While a cathode with a planar emission surface may perform adequately under certain conditions, issues can arise during transmission under a radial focusing

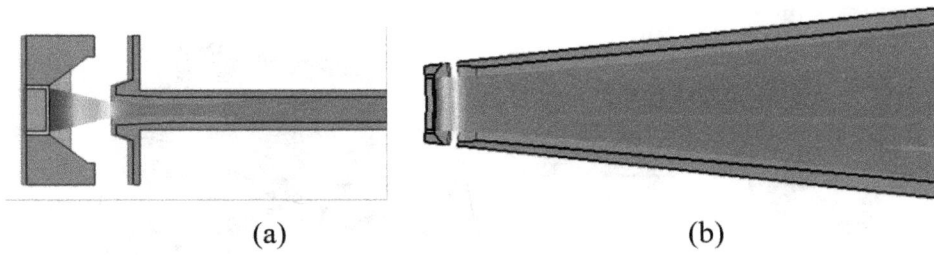

(a)   (b)

Figure 3.19. Electron trajectory of a cylindrical emission surface in the absence of a magnetic field. (a) Side view; (b) top view. © [2015] IEEE. Reprinted, with permission, from [11].

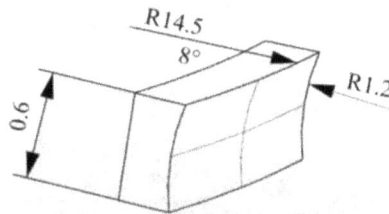

Figure 3.20. Cathode with a doubly curved emission surface (units: mm).

magnetic field due to insufficient lamination. A potential solution involves replacing the planar emission surface with a doubly curved emission surface. This surface is characterized by the feature that the plane parallel to the z-axis intersects the hyperboloid in the form of a hyperbola (or a pair of intersecting lines). In this context, a hyperbola is defined as the locus of a moving point where the difference of the distances from two fixed points (the foci) remains constant, equivalent to the real axis $2a$. The hyperboloid cathode emission surface described in this context may not strictly adhere to the mathematical definition of a hyperboloid, as illustrated in figure 3.20. The use of a hyperboloid emission surface can increase the emission surface area of a cathode while retaining the same angle and height. This design enhancement enables the cathode to reach a greater emission current under identical working conditions.

Figure 3.21 illustrates the trajectory of an electron beam originating from a cathode featuring a doubly curved emission surface in the absence of a magnetic field. The axial fluctuation of the electron beam surpasses that observed for a cylindrical emission surface. Nevertheless, the electron beam can still attain a 100% flow rate.

3.3.2 Radial tunable PCM magnetic focusing system

For a miniaturized planar TWT with an angular logarithmic microstrip meander-line SWS, simple disc-shaped coil or permanent magnets can provide a radial focusing magnetic field, but they fail to meet the requirements of miniaturization. In addition, to prevent the penetration of the magnetic field into the electron gun area, a circular pole piece must be incorporated into the inner circle section of the disc-

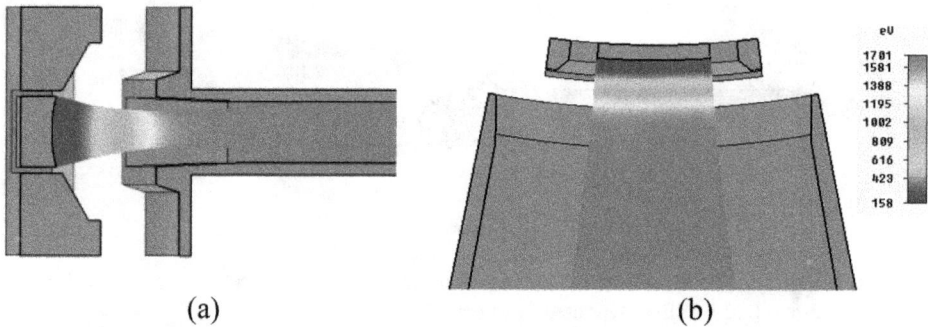

Figure 3.21. Electron beam trajectory produced by a hyperbolic emission surface (in the absence of a magnetic field). (a) Axial cross-sectional view; (b) angular cutaway view.

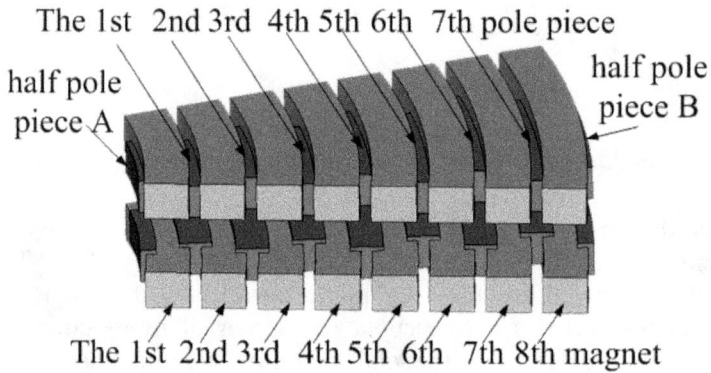

Figure 3.22. Simulation model of a radially tunable PCM focusing system. © [2017] IEEE. Reprinted, with permission, from [12].

shaped magnetic focusing structure. This requirement poses a challenge due to the conflict with the flanged magnetic screen utilized in the current TWT flow process. This conflict introduces complexities into the overall tube assembly design.

The utilization of a tunable PCM magnetic focusing system proves to be a viable solution for managing the angular radial sheet electron beam, as depicted in figure 3.22. The focusing system comprises four periods, each characterized by two pairs of axially symmetric fan-shaped magnetic blocks arranged both above and below the beam; their separation is facilitated by angular sliding pole pieces. Half-pole piece A and half-pole piece B are arranged at the start and end positions.

The structural dimensions are shown in figure 3.23. The pole pieces have a 'T' shape, and their dimensions are specified as follows: the thickness of the pole piece is denoted by L_{p1}, the height of the pole piece is H_{p2}, and the width and thickness of the pole piece foot are L_{p2} and H_{p1}, respectively. The radius of the small arc surface of the pole piece is denoted by R_{pai} and the radius of the small arc surface of the pole piece foot is R_{pbi}. The thickness of the angular magnetic block is H_m, its radial length is L_m, and the radius of the small arc surface is R_{mi}.

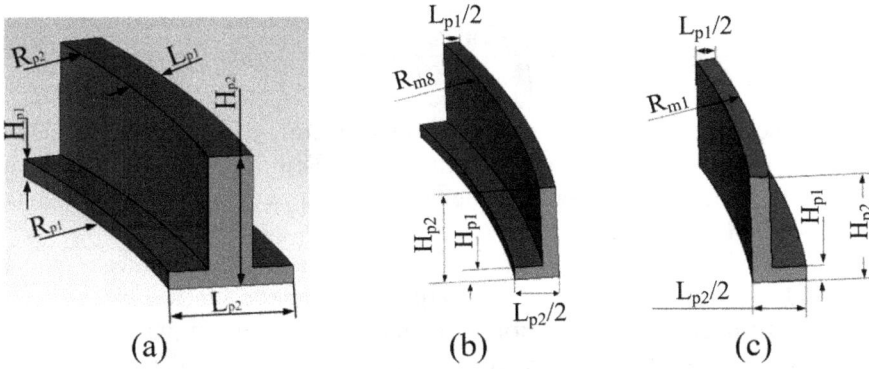

Figure 3.23. Structural dimensions of the pole pieces. (a) Pole piece; (b) half-pole piece A; (c) half-pole piece B. © [2017] IEEE. Reprinted, with permission, from [12].

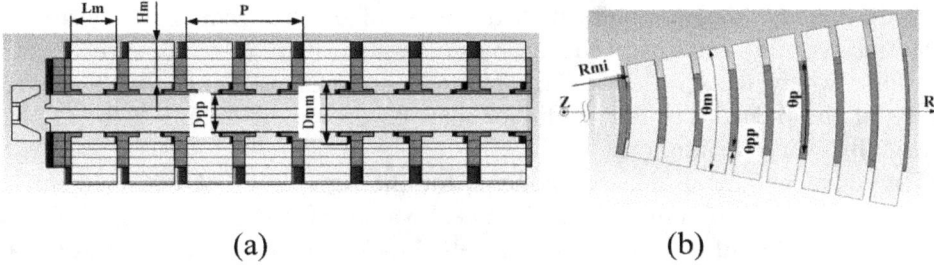

Figure 3.24. Assembly of a radially tunable PCM magnetic focusing system. (a) Central axis section; (b) axial top view. © [2017] IEEE. Reprinted, with permission, from [12].

The assembly diagram of a radially tunable PCM magnetic focusing system is shown in figure 3.24. The angles of the fan-shaped magnetic block and the pole piece are θ_m and θ_p, respectively, while θ_{pp} is the cross angle between the upper and lower pole pieces in the same radial position. The specific settings are as follows: in the top axial view (figure 3.24(b)), angles of clockwise angular rotation are defined as positive. The angular rotation of half-pole piece A in the positive axial half-axis is $-\theta_{pp}/2$, and the angular rotation of half-pole piece A in the negative axial half-axis is $\theta_{pp}/2$. In this way, a staggered angle of θ_{pp} is formed between the upper and lower half-pole pieces A. In the direction of increasing radius, the upper and lower pole pieces of the first pair are rotated through angles of $\theta_{pp}/2$ and $-\theta_{pp}/2$, respectively, and the subsequent pole pieces are alternately rotated by the same amounts. The Brillouin magnetic field and the period of the PCM focusing field are given by equations (3.13) and (3.14), respectively:

$$B_b = \sqrt{\frac{\sqrt{2}\,I_0}{4wt\gamma\varepsilon_0\eta^{3/2}V_0^{1/2}}} \qquad (3.13)$$

$$P \approx \pi \sqrt{\frac{\sqrt{2}\,wt\varepsilon_0 V_0^{3/2}\eta^{1/2}}{2I_0}}. \tag{3.14}$$

Given the electrical parameters of the electron beam, it is possible to determine reference values for key parameters, including the Brillouin magnetic field and the period length of the PCM magnetic focusing system. Table 3.4 lists the structural parameters.

The values of the Brillouin focusing magnetic field obtained using the radial uniform and the periodic magnetic field equations are shown in table 3.5.

In the section concerning periodic magnetic field design within the 'Design Manual of Small and Medium Power Traveling-Wave Tube (Trial version)', it is noted that the magnetic field peak of the first magnetic block in the focusing system should be 1.5 to 2 times the value of the Brillouin magnetic field. This consideration arises from several factors, including the inability of the cathode region to fully shield the magnetic field, variations in the electron gun, the impact of the electrons' thermal velocity, the nonuniform distribution of electron emission density, and potential assembly errors introduced by grid usage or other factors. However, in situations where the envelope of the electron beam closely aligns with the inner radius of the pole piece, the multiplier may be decreased, falling below 1.5 and potentially even reaching one.

The radial magnetic field components for the tunable PCM obtained through simulation are shown in figure 3.25. The peak values of the radial magnetic field for the first, second, and third pairs of magnetic blocks are shown to occur at 633 Gs,

Table 3.4. The structural parameters of the radially tunable PCM magnetic focusing system.

θ_m	θ_p	θ_{pp}	R_{mi}		H_m	L_m
22°	13°	0.9°	$16.65 + 2.4 \times (i-1)$ $i = 1,2,3,4,5,6,7,8$		1.7 mm	1.9 mm

R_{pai}	R_{pbi}	L_{p1}	L_{p2}	H_{p1}	H_{p2}	P
$16.65 + 2.4 \times i$ (mm) $i = 1,2,3,4,5,6,7$	R_{pai} −0.45 (mm)	0.5 mm	1.4 mm	0.2 mm	1.5 mm	4.8 mm

Table 3.5. A comparison of the calculated values for two kinds of Brillouin magnetic fields.

	V_0 (V)	I_0 (A)	t (mm)	r (mm)	θ (°)	B_b (Gs)
Radial Brillouin focusing magnetic field (periodic)	1700	0.0512	0.26	$w \approx r\theta\pi/180$	8	335.5
Radial Brillouin focusing magnetic field (uniform)	1700	0.0512	$a_0 = t/2$	16.5	8	581.5

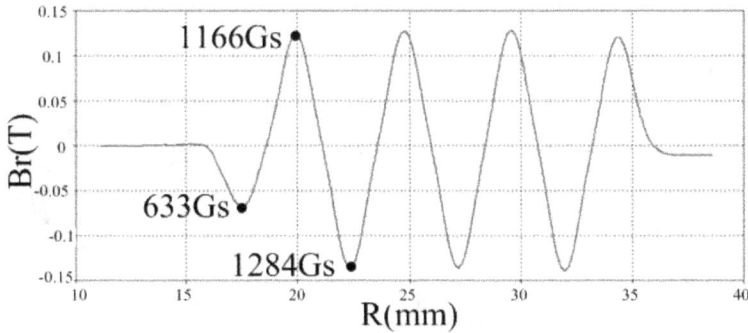

Figure 3.25. Radial component of the radial tunable PCM focusing system. © [2017] IEEE. Reprinted, with permission, from [12].

1166 Gs, and 1284 Gs, respectively. The ratio between the peak value of the radial magnetic field for the first pair of magnetic blocks and the Brillouin magnetic field value was calculated using the two different methods listed in table 3.5 and found to be 1.89 and 1.09, respectively. This signifies that the radial magnetic field peak value for the first pair of magnetic blocks is one to two times the Brillouin magnetic field value. As such, the magnitude of the magnetic field is established, and it can be concluded that the influence of the angular magnetic field component on the motion of the electron beam can be neglected.

The description and results regarding the magnetic field values and Brillouin magnetic field values for the aforementioned magnetic system can be acquired through particle motion simulation. The angular radial electron beam exhibits a flow rate of 100%, indicating effective preservation of radial motion. Consequently, the theoretical guidance provided by the Brillouin magnetic field calculation is substantiated in a preliminary manner. The constraints imposed by the radially tunable PCM magnetic focusing system on fan-shaped electron beams are influenced by various factors, including the rotational angle of the pole piece within the magnetic focusing system and the period length of the magnetic system.

Figure 3.26 illustrates the effect of the rotational angle ($\theta_r = \pm\theta_{pp}/2$) on the axial component of a radially tunable PCM focusing system. The positive and negative symbols associated with θ_r at the sampling positions in the radial position $r = 26$ mm indicate the rotation direction. Specifically, in the axial top view, a clockwise rotational angle of the half-pole piece A is denoted by a positive value. In the central region, the axial component of the magnetic field is observed to be zero, while at the angular edges, where the angles are staggered, an axial (z) magnetic field is present. When the rotational angle transitions from negative to positive, the axial component of the magnetic field changes direction, and its magnitude varies from several Gauss to tens of Gauss in response to the angular adjustment. It is noteworthy that the angles exhibit symmetric centrality on both sides. In addition, as the rotational angle increases, the value of the axial magnetic field component at the edge becomes more pronounced.

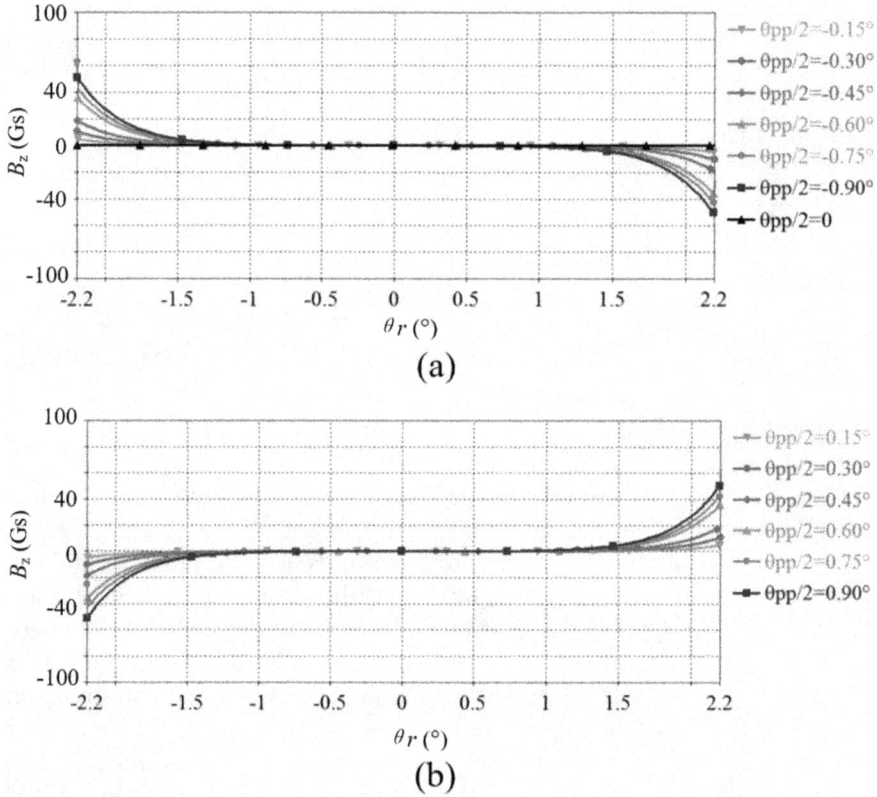

Figure 3.26. Influence of pole piece rotational angle on the axial component of the magnetic field. (a) Distribution curve of $\theta_{pp} \leq 0$; (b) distribution curve of $\theta_{pp} > 0$. © [2017] IEEE. Reprinted, with permission, from [12].

Introducing an axial magnetic field into the angular edge of a fan-shaped radial electron beam mitigates or eliminates the diocotron instability. To examine the distribution of the axial magnetic field in a radially tunable PCM magnetic focusing system, the axial magnetic field component was investigated at a rotational angle of $-0.45°$, as depicted in figure 3.27. In the central region, the axial magnetic field remained negligible, with a magnitude on the order of 10^{-7} T. However, at the angular edge, the magnetic field axis was distributed in the tunnel region and had values ranging from 10 to 20 Gs. Notably, the field amplitudes along the two sampling lines were identical but exhibited opposite polarities.

The assembly of the angular radial tunable PCM with the electron gun is shown in figure 3.28. This assembly fulfills the following two criteria: (a) angular central section coincidence; (b) the concentricity of the radial tunable PCM magnetic block, the pole-piece, and the electron gun electrode.

The operating parameters are set as follows: anode voltage 1700 V, emission current 51.2 mA, current emission density and perveance of the corresponding cathode were 4.21 A cm^{-2} and 0.73 e-6 P, respectively. In this section, the primary

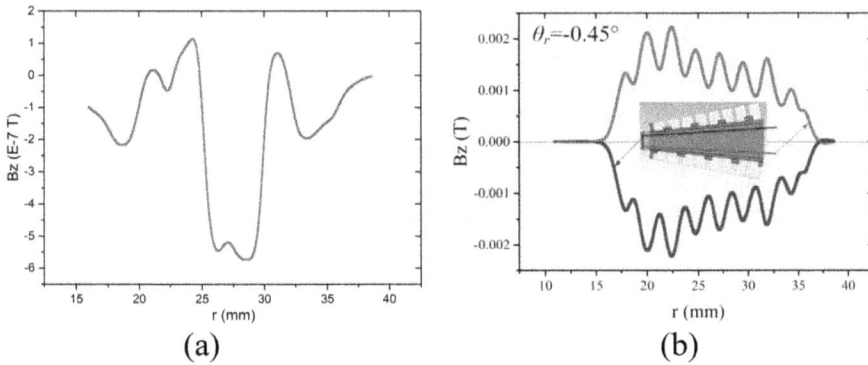

Figure 3.27. Distribution of the axial component of the magnetic field at a rotational angle of the pole piece of $-0.45°$. (a) Distribution curve of the central region; (b) distribution curve at the angular edge. © [2017] IEEE. Reprinted, with permission, from [12].

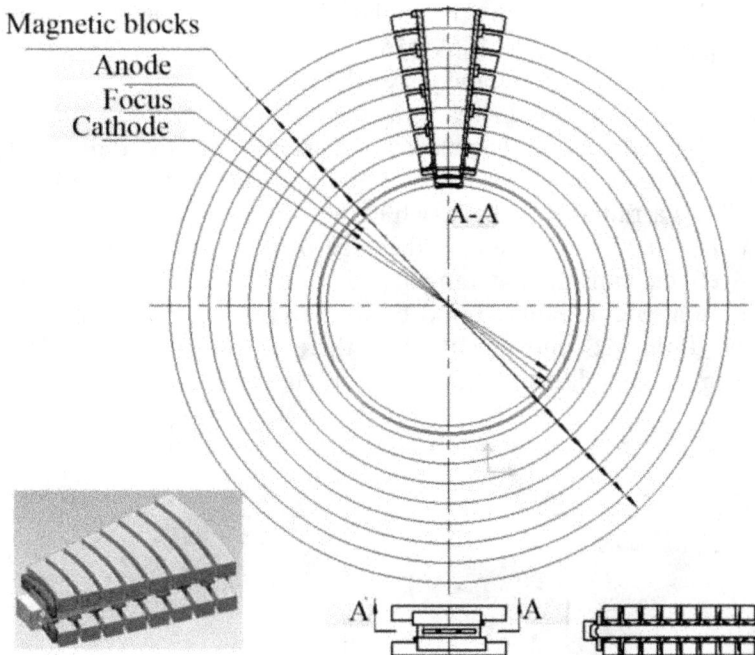

Figure 3.28. Assembly of the angular radial tunable PCM focusing system and angular radial sheet beam electron gun.

focus is to study the impacts of the pole-piece rotational angle and period length of the radially tunable PCM magnetic focusing system on the flow rate. This emphasis is due to practical considerations in the testing process, where, during the processing of the magnetic system with a specific period length, only the rotational angle of the pole piece can be adjusted.

Figure 3.29(a) displays the relationship curves for the rotational angle versus the flow rate. When the rotational angle is positive, the flow rate decreases sharply as the rotational angle increases. When the rotational angle is negative, the flow rate remains greater period than 98%. Notably, at rotational angles of −0.3°, −0.4°, and −0.5°, the system achieves a 100% circulation rate. While the flow rate remains very high at other angles, it is observed that the trajectory of the electron beam becomes distorted. Therefore, the optimal choice of rotational angle lies within the range of −0.3° to −0.5°. Figure 3.29(b) shows the relationship curve between the period and the flow rate of the radially tunable PCM focusing system. The horizontal axis parameter in the figure represents half-period data. Half periods with a flow rate of 100% have lengths of 2.2 mm, 2.3 mm, and 2.4 mm. The flow rate experiences a rapid decline when the half-period value falls below or above this specified range. To alleviate the challenges associated with processing and assembly, a principal selection involves opting for larger period values.

The electron distribution on the radial section of the electron gun region is shown in figure 3.30(a). As electrons leave the emission surface, the axial thickness of the beam progressively diminishes as the radial distance increases until the maximum compression position is reached. At 16.5 mm, the electron beam thickness is approximately 0.23 mm. Simultaneously, the electron beam of the x-component of the rectangular x–y section increases as the radial distance expands, indicating a divergent trend in its motion. The electron distribution in the radial section of the beam tunnel region is shown in figure 3.30(b). In the tunnel, the electron beam primarily follows a radial trajectory, while the y-component fluctuates in the state distribution. The axial component at the edge of the focusing system introduces a constraining effect, causing the fan-shaped radially divergent electron beam to fluctuate within a specific range. Importantly, the beam remains unimpeded by the channel wall during its movement, and there is no significant divergence or convergence in the angular direction. Even as the trajectory of the electron beam

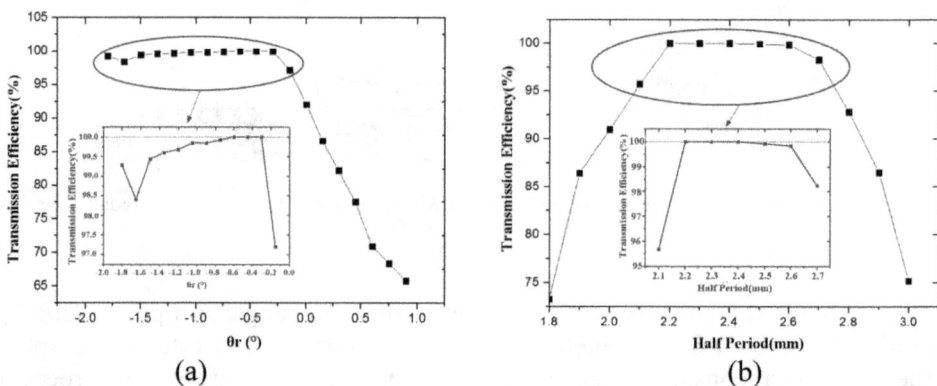

Figure 3.29. Curves showing the relation between (a) rotational angle and flow rate; (b) period and flow rate of the angular radial tunable PCM focusing system. © [2017] IEEE. Reprinted, with permission, from [12].

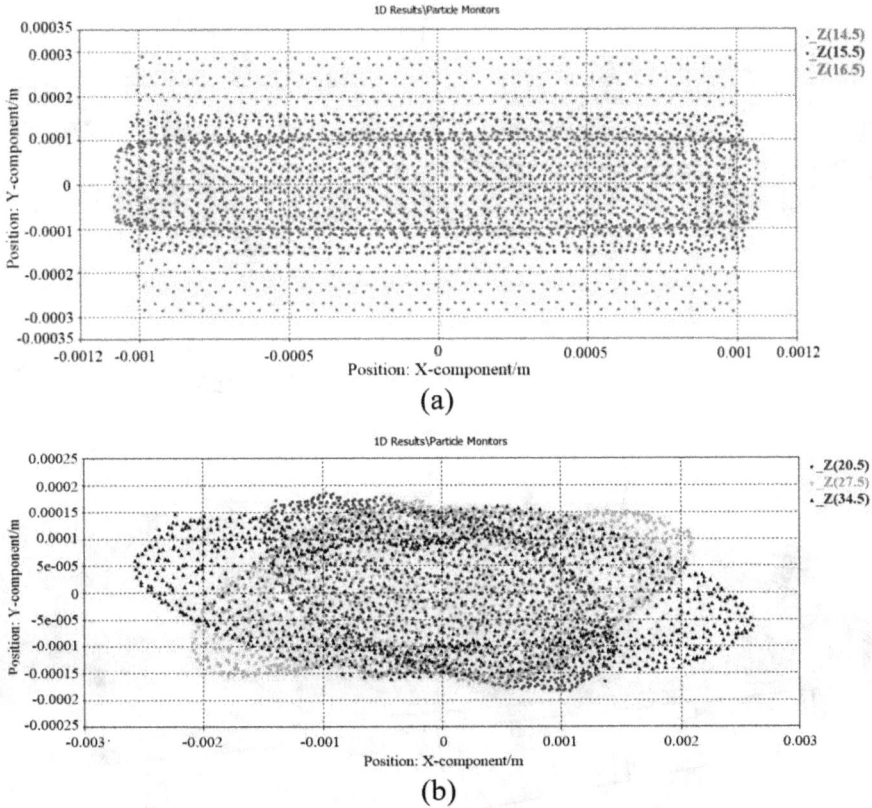

Figure 3.30. Distribution of the electron beam in different radial sections: (a) gun area: $r = 14.5$ mm, 15.5 mm, 16.5 mm; (b) tunnel area: $r = 20.5$ mm, 27.5 mm, 34.5 mm. © [2017] IEEE. Reprinted, with permission, from [12].

maintains its fan-shaped form, it does not exhibit substantial angular spreading or convergence.

Figure 3.31 shows a sectional diagram of the electron trajectory within the angular radial tunable PCM focusing system. Notably, the entire radially divergent sheet electron beam, accounting for 100% of the electrons, successfully traverses the angular beam tunnel for a distance of 20 mm. In addition, the duty ratio is approximately 50%.

The simulation results for the angular radial electron optical system affirm that the radially tunable PCM focusing system effectively constrains the sector radially divergent strip electron beam focusing. Its advantages, including its compact size and the absence of interference between the assembly and the electron gun shell components, address challenges associated with the large volume and intricate assembly requirements encountered when utilizing a disc-shaped uniform magnetic field with an angular radial electron beam.

→ magnetizing direction ——→ magnetic line

(a)

(b)

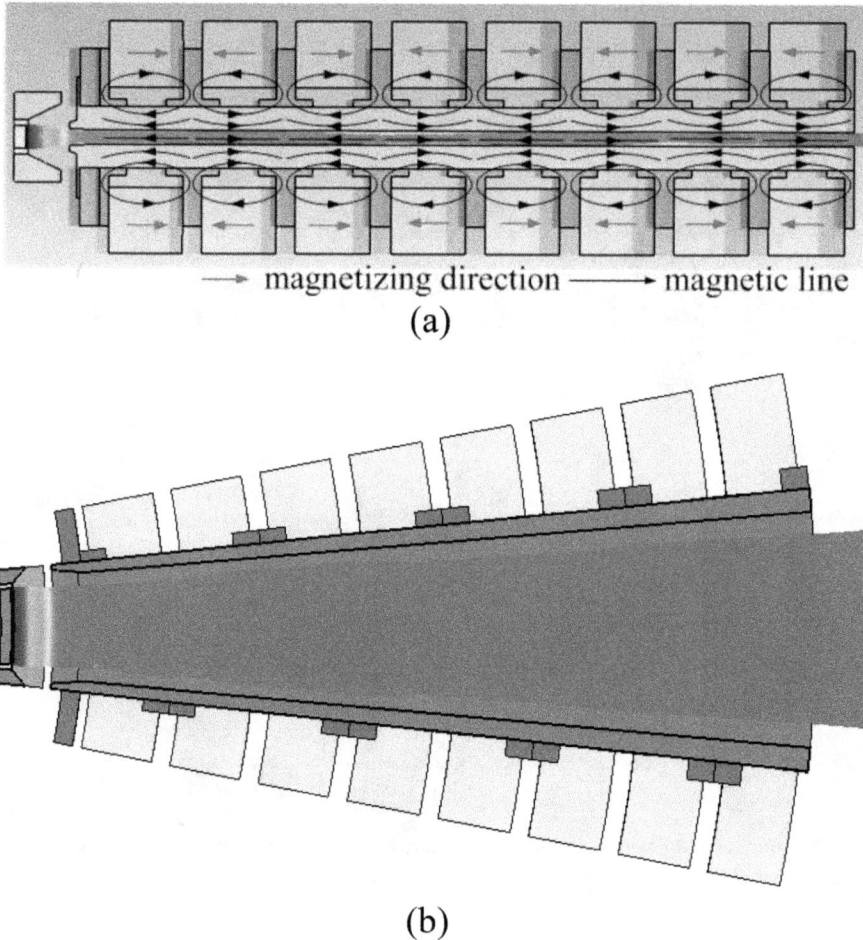

Figure 3.31. Electron beam trajectory used in co-simulation to determine the structural parameters. (a) Angular central section; (b) axial central section. © [2017] IEEE. Reprinted, with permission, from [12].

3.4 Sheet electron beam EOSs

The conventional sheet electron beam was introduced in the 1990s in conjunction with the planar SWS. This type of electron beam is characterized by a rectangular or elliptical cross-sectional area and retains its shape in the transmission direction. The key aspects to consider for the sheet electron beam are its formation and its transmission.

3.4.1 Sheet beam electron gun

In the sheet beam EOS, a cathode with a circular, elliptical, or rectangular surface is typically utilized to emit electrons. The electric field produced between the beam-forming electrode and the anode plays a crucial role in gradually guiding and compressing the emitted electrons into a sheet-like form. Simultaneously, as electrons enter the anode channel, they are accelerated to their designated energy levels. The magnitude of the generated current and the cross-sectional area of the sheet are influenced by factors such as the shape of the beam-forming electrode and anode nose, as well as the distance between the anode and the cathode.

The design of a sheet beam electron gun starts with the electric parameters determined by the matched planar SWSs. These parameters typically include the operating voltage, emitted current, and cross-sectional size. The design approach adopted here is based upon an investigation of the beam–wave interaction of a W-band meander strip-line SWS, whose cross-sectional sheet beam size is specified as 0.6 mm × 0.1 mm and whose voltage and current are set to 11.3 kV and 0.1 A, respectively; the bottom of the sheet beam is positioned 0.05 mm from the surface of the strip line, and the equivalent electron beam tunnel height is 0.2 mm. To generate a sheet beam that matches the required shape, size, and working parameters, we use the CST software Particle-Tracking E-static Solver to design an electron gun that employs a circular cathode to compress and generate the sheet beam.

Figure 3.32 illustrates the sheet-beam electron gun, which is designed in accordance with the Pierce gun theory. This design primarily consists of three key components: a circular cathode, a beam-forming electrode featuring an elliptical cross-section, and an anode nose with an elliptical opening. The beam-forming electrode and cathode are connected to a negative high voltage of −11.3 kV. In this configuration, the x and y openings of the beam-forming electrode are larger than those of the anode nose, establishing a gradually converging electric field between the beam-forming electrode and the anode. The distributions of the electric field and the potential are depicted in figure 3.33. The desired size and current of the sheet electron beam are obtained by iteratively optimizing the structural size and spacing of the beam-forming electrode and anode nose. After 29 iterations, a current of 0.112 A is finally predicted.

Meanwhile, figure 3.34 shows the trajectory envelope and the particle distribution in the cross-sectional area. The electron beam emitted by the circular cathode with a

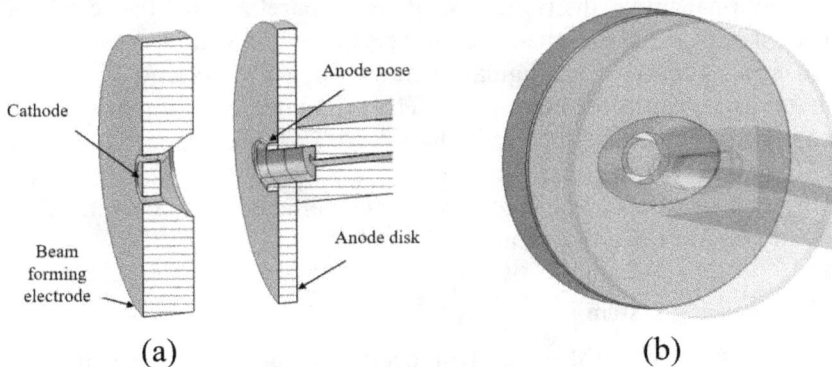

Figure 3.32. Schematic diagram of the electron gun structure. (a) Cross-sectional profile; (b) perspective view.

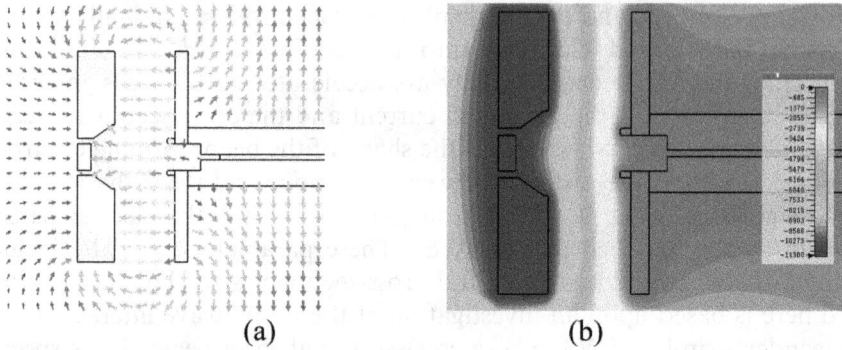

Figure 3.33. Electric field distribution of the electron gun. (a) Diagram of the electric field distribution; (b) a potential distribution diagram.

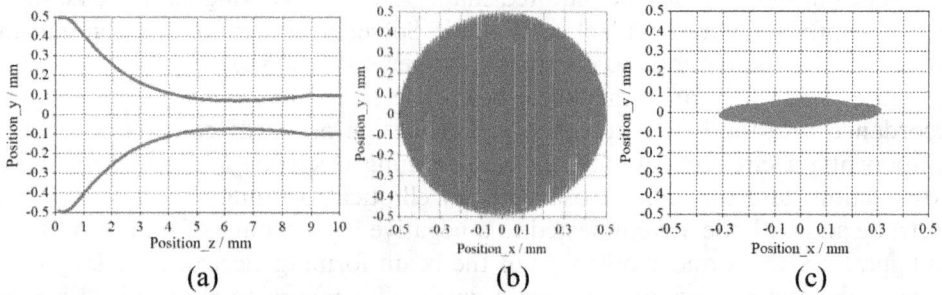

Figure 3.34. Electron beam trajectory. (a) Trajectory envelope in the y-direction; (b) transverse distribution at the cathode emission surface; (c) transverse distribution at the waist. Copyright (2024) IEEE. Reprinted, with permission, from [15].

radius of 0.5 mm is compressed rapidly in the y-direction, reaching its maximum compression at $z = 6.4$ mm. Subsequently, the beam gradually expands in the specified direction. Therefore, the point $z = 6.4$ mm is referred to as the beam waist, which is commonly used as the starting point of the focusing magnetic field. From the cross-sectional area distributions, it is apparent that the electron beam experiences significant simultaneous compression in the x- and y-directions, resulting in an approximately rectangular beam with a cross-section of 0.62 mm × 0.14 mm (an aspect ratio of about 4.4:1). The calculated emission current density at the cathode surface is determined to be 14.1 A cm^{-2}, while the current density at the strip located at the beam waist is approximately 126.7 A cm^{-2}. This results in an areal compression ratio of about 9:1. As the current remains constant, the current density compression ratio is also approximately 9:1.

3.4.2 PCM focusing system

In the context of focusing a sheet electron beam, a primary challenge arises from the presence of the $E \times B$ shear force, which induces rotation of the beam around the axis of symmetry. Due to the periodic interleaving of the pole piece in the x-direction, the PCM magnetic focusing system can generate an additional magnetic field to control the rotation of the left and right ends of the sheet beam in the y-direction. The B_y produced by periodic positive and negative transformation ensures that the sheet beam rotates clockwise and counterclockwise, thereby maintaining the envelope of the sheet beam within the range of the electron beam tunnel.

The assembly of the strip-line SWS, electron gun, and PCM magnetic system is shown in figure 3.35. The permanent magnets used in this system are typically constructed from materials such as samarium cobalt alloys or NdFeB. Due to its high remanence (usually up to 1.2 T) and high coercivity, NdFeB is known as the magnetic 'king'. The pole piece is made of iron. To prevent the magnetic field from penetrating the electron gun area, a magnetic shield made of iron is set up to block the magnetic field.

The structural dimensions of the designed magnet and the pole piece are shown in figure 3.36, where the parameter 'cuo' represents the offset of the pole piece in the x-direction.

The amplitude and period length of the axial magnetic field B_z are generally calculated from the Brillouin magnetic field B_b and the plasma wavelength λ_p to give a reference range using the following formulas:

$$B_b = \sqrt{\frac{\sqrt{2}\,I_0}{4wt\gamma\varepsilon_0\eta^{\frac{3}{2}}V_0^{\frac{1}{2}}}} \tag{3.15}$$

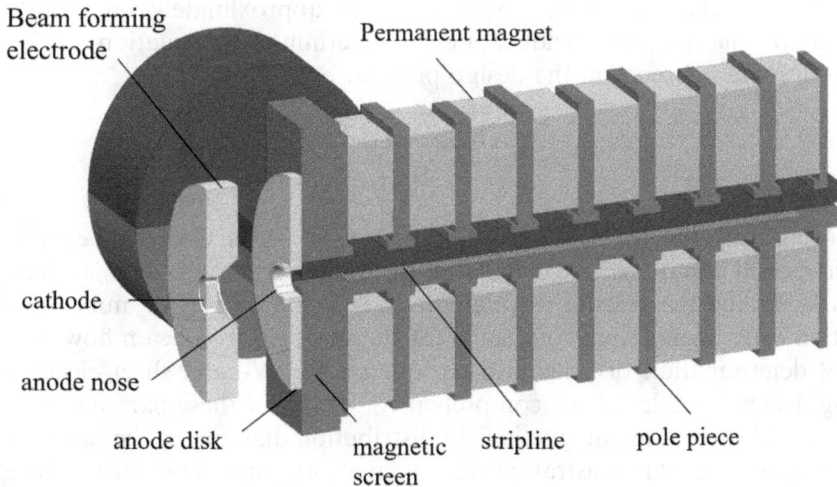

Beam forming electrode
Permanent magnet
cathode
anode nose
anode disk
magnetic screen
stripline
pole piece

Figure 3.35. Schematic diagram of the structural model of the W-band EOS. Copyright (2024) IEEE. Reprinted, with permission, from [15].

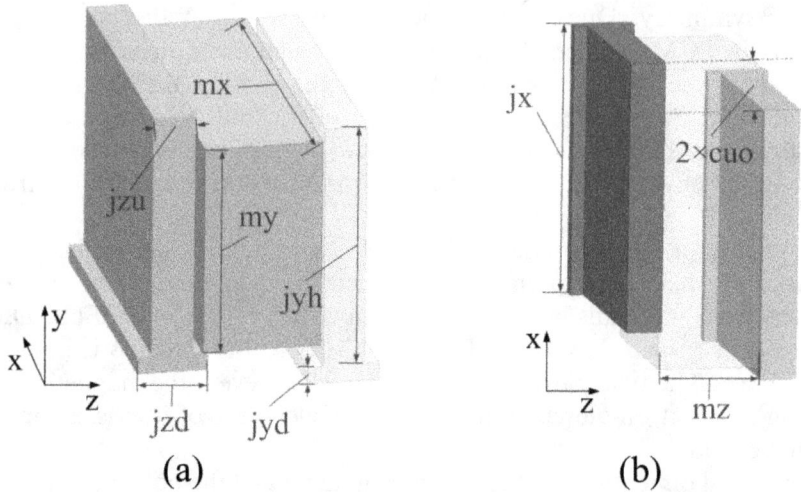

Figure 3.36. Dimensions of the magnet and the pole piece. (a) Three-dimensional view; (b) an x–z-plane view.

$$\lambda_p = \frac{2\pi v}{\omega_p} = \frac{2\pi v}{\sqrt{\frac{eJ_0}{m\varepsilon_0 v}}} = \frac{2\pi v}{\sqrt{\frac{\eta I_0}{\varepsilon_0 w t v}}} \tag{3.16}$$

wherein V_0 and I_0 are the sheet electron beam voltage and current, respectively; w and t are the sheet electron beam width and thickness, respectively; ε_0 and η are the electric constant of vacuum and the charge mass ratio, respectively; and γ is the relativistic factor. In accordance with established engineering principles, the required axial magnetic field B_z is set to approximately 1.5–2 times the value of the Brillouin magnetic field, and the PCM length is set to approximately one third of λ_p. Furthermore, the magnetic field period, as determined by equation (3.17), serves as an additional reference in the design process:

$$P \approx \pi \sqrt{\frac{\sqrt{2}\, w t \varepsilon_0 \, V_0^{\frac{3}{2}} \eta^{\frac{1}{2}}}{2 I_0}}. \tag{3.17}$$

By incorporating the relevant information for the sheet electron beam into the aforementioned equations, the calculated Brillouin magnetic field B_b is determined to be 0.16 T, and the plasma wavelength λ_p is found to be 21.2 mm. Subsequent simulation experiments were conducted for the sheet electron beam flow, leading to the final determination of structural parameters for W-band sheet electron beam focusing. Refer to table 3.6 for comprehensive details of these parameters.

Figure 3.37 depicts the magnetic field distribution diagram of the designed PCM focusing system. In this illustration, the magnets are organized into eight groups. The first group consists of two magnets, one upper and one lower, both sharing the same magnetization direction and size, and they are magnetized at 0.3 T. The second group is magnetized at 0.6 T, while the subsequent six groups are all magnetized

Table 3.6. The structure and size of magnet and pole piece used for the sheet electron beam PCM focusing.

Parameter	Value (mm)	Parameter	Value (mm)
jx	7	mx	7.6
jyd	0.3	my	3.6
jyh	4.2	mz	2
jzd	1.2	cuo	0.5
jzu	0.7	/	/

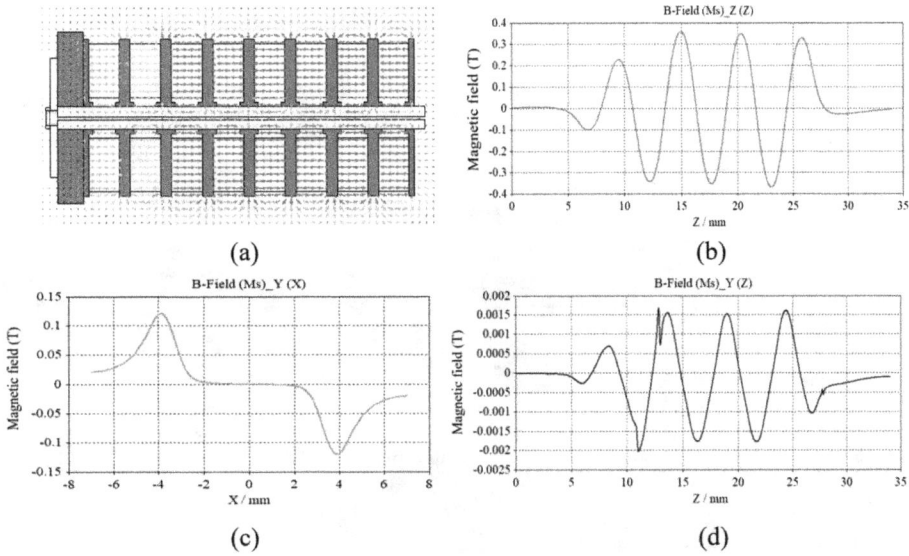

(a)

(b)

(c)

(d)

Figure 3.37. Magnetic field distribution of the PCM focusing system for a sheet electron beam. (a) Magnetic field intensity curve; (b) B_z component in the z-direction at $x = 0$; (c) B_y component in the x-direction at $z = 15$ mm; (d) B_y component in the z-direction at $x = 0.3$ mm (beam edge).

at 1 T. Notably, each of these eight groups exhibits an alternate magnetization direction. As evident from the figure, an axial magnetic field is introduced into the electron beam channel using the guidance of an iron pole-piece pin. The peak value of B_z is approximately 0.35 T. The distribution of the magnetic field component B_y is symmetrically centered, generated by the staggered arrangement of the pole pieces. At the edge of the sheet electron beam, the peak value of the B_y component is around 15 Gs, a level sufficient to effectively control the rotation at the periphery of the sheet electron beam.

The flow of the sheet beam can be investigated by a co-modulation simulation experiment, which is conducted using the W-band sheet beam gun and the matching PCM focusing system. The simulated trajectory of the sheet electron beam is illustrated in figure 3.38, which reveals noticeable compression in both the broad and narrow edges of the beam.

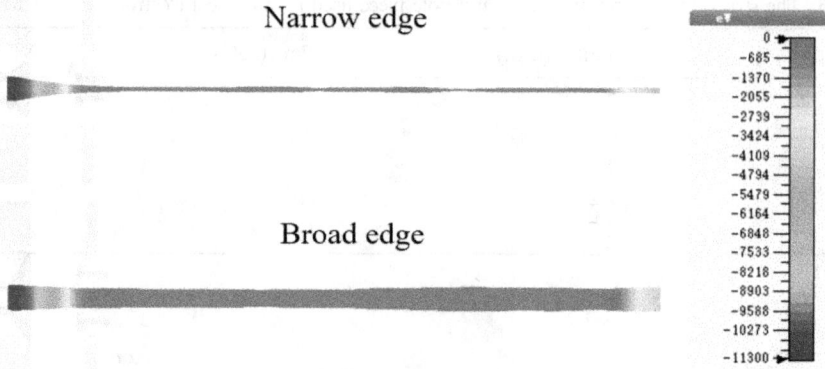

Figure 3.38. Sheet beam compression in the narrow and wide edges.

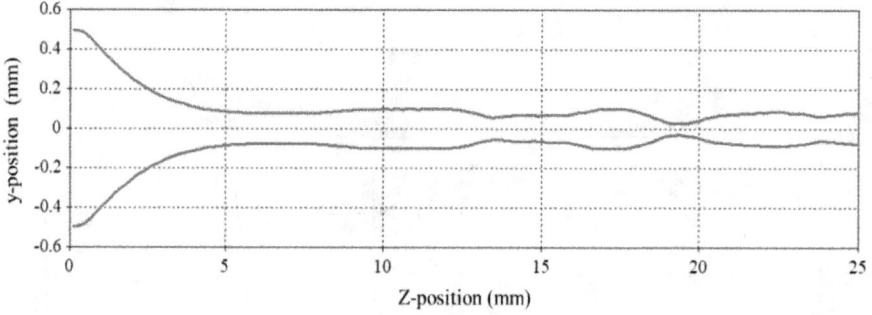

(a)

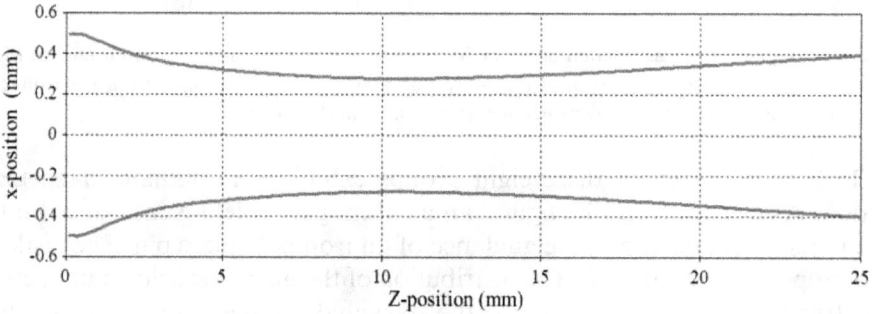

(b)

Figure 3.39. Electron beam envelopes: (a) narrow-edge fluctuations; (b) broad-edge fluctuations.

The beam envelope, as depicted in figure 3.39, provides a quantitative representation of the fluctuations in the wide and narrow sides of the beam during transmission. At the beam waist, the broad edge of the electron beam is compressed to the designated value of 0.6 mm. Subsequently, the beam envelope gradually

Table 3.7. The collector currents of the electron beam at different voltages.

Cathode voltage (V)	Interception current (A)	Collector current (A)	Flow rate
−10 300	0.001 591	0.096 076	98.3712%
−10 800	0.002 157	0.102 704	97.9429%
−11 300	0.002 562	0.109 667	97.7174%
−11 800	0.002 804	0.116 954	97.6589%
−12 300	0.002 94	0.124 505	97.6928%

expands due to the absence of a focusing force. On the narrow side, the B_y generated by the pole piece interlace controls the swing amplitude of the electron beam in the y-direction. This ensures that a significant portion of the envelope on the narrow side remains within the electron beam channel at a height of 0.2 mm. The final simulation results indicate an impressive flow rate of 97.7% for the 20 mm long electronic beam tunnel.

The impact of beam voltage on the flow rate is evident, as summarized in table 3.7, which outlines the emission current and final flow rate under various operating voltages. With a gradual increase in the working voltage, the emission current also rises, and although there is a slight decrease in flow rate, all of which exceed 97%. These findings serve as a valuable reference for the design and performance considerations of the TWT.

3.5 Multiple sheet beam EOSs

The utilization of multiple cylindrical-beam EOSs has become prevalent in high-power applications, such as higher-harmonic klystrons and TWTs. This approach is adopted due to its ability to deliver currents significantly higher than those achievable with a single cylindrical beam. In line with the integration motivation underlying planar SWSs, the study of multiple sheet beam EOSs has great importance.

3.5.1 Multiple angular radial sheet beam EOSs

The angular cascaded angular log SWS facilitates beam–wave interaction with electromagnetic waves through the use of double angular radial electron beams. Therefore, it is imperative to design a corresponding EOS tailored for dual electron beams. Such a design should ensure that there is no interference between the radial twin electron beams during the beam–wave interaction [13].

3.5.1.1 Electron gun
This section introduces the design and investigation of a radial double-beam electron gun. The primary function of the radial double-beam electron gun is to transmit and compress the radially diverging double beam. Figure 3.40 illustrates the structure of a radial dual-beam electron gun. The electron gun features a hot cathode based on electron thermal emission, constructed using tungsten as the base material. The

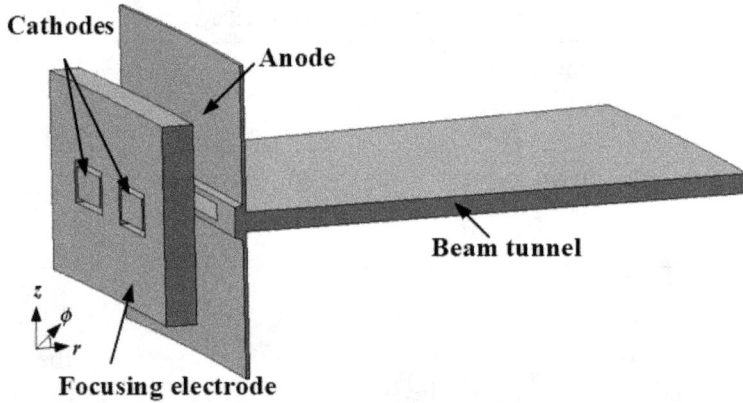

Figure 3.40. Schematic diagram of a radial dual-beam electron gun. Reproduced from [13]. CC BY 4.0.

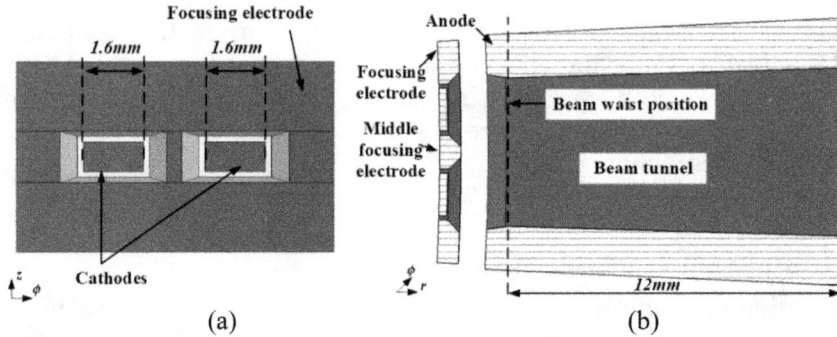

(a)　　　　　　　　　　　　　　　(b)

Figure 3.41. Schematic diagram of the relative position of the radial dual-beam electron gun. (a) Schematic diagram of cathode and focusing electrode structure. (b) Relative positions of cathode, focusing electrode, and anode. Reproduced from [13]. CC BY 4.0.

electron emission layer employs a salt-impregnated binding layer. Electrons are emitted from a fan-shaped surface on two cathodes, and they are subsequently compressed and accelerated by a set of focusing electrodes and anodes. The positions of the cathode, focusing electrode, and anode are detailed in figure 3.41. In scenarios where the electron gun emits a single electron beam, the angular divergence of the radial electron beam results in a reduced compression ratio in the φ-direction (i.e. the angular direction). Consequently, only focused compression in the axial z-direction is necessary for the electron beam. However, when the electron gun generates a double electron beam, the two cathodes, being in close proximity, share the same potential. This alters the electric field between the two cathode regions, leading to deflection of the electron beam during the emission phase. Hence, special attention is required to achieve focusing in the φ-direction within the electron gun region.

　　This section introduces the design of the middle focusing pole, whose purpose is to ensure a well-defined trajectory for the double electron beam within the electron gun region. The proposed design incorporates potential compensation for the middle focusing pole, which contributes to achieving a symmetrical electric field distribution in the electron gun region. This symmetry is crucial for ensuring the ideal trajectory of the electron beam.

Figure 3.42 illustrates the distribution of the electric field isolines and electric field intensity in the region between the anode and cathode, comparing the electric field in the presence and absence of an intermediate-focus electron gun. In the absence of intermediate focus, a potential depression is observed in the area between the two cathodes, as depicted in figure 3.42(a). This configuration results in an 'inward' electric field acting on the electron beam. The region of the two electron beams close to the 'inward' electric field experiences an inward electric field force, overpowering the outward coulomb repulsion force between the two electron beams. Consequently, as shown in figure 3.43(a), the electron beams merge after advancing a certain distance.

When an intermediate focusing pole is added between the two cathodes, it causes an 'outward' electric field to act on the electrons in the central region. The electric field distribution experienced by the electron beam is completely symmetric with

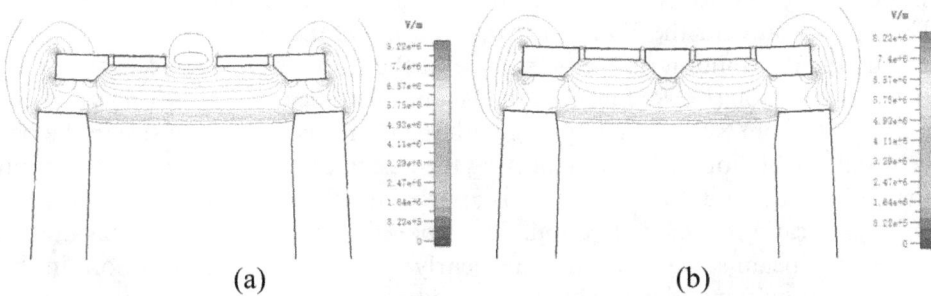

Figure 3.42. Equipotential distribution of the electric field. (a) Electron guns without intermediate focusing electrodes. (b) Electron guns with an intermediate focusing electrode. Reproduced from [13]. CC BY 4.0.

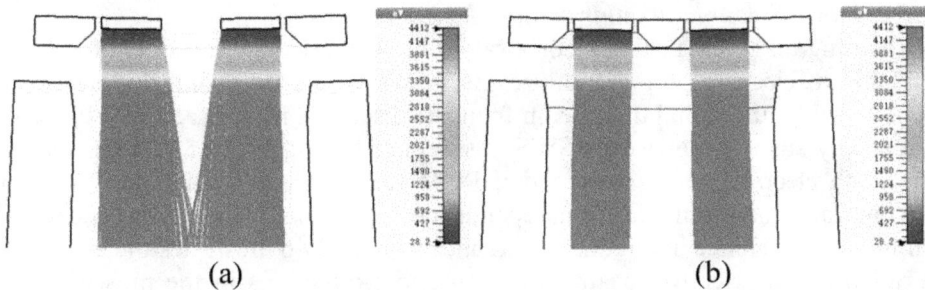

Figure 3.43. Schematic diagrams of the electron beams. (a) Trajectories in the absence of intermediate focusing poles. (b) Trajectories with intermediate focusing poles. Reproduced from [13]. CC BY 4.0.

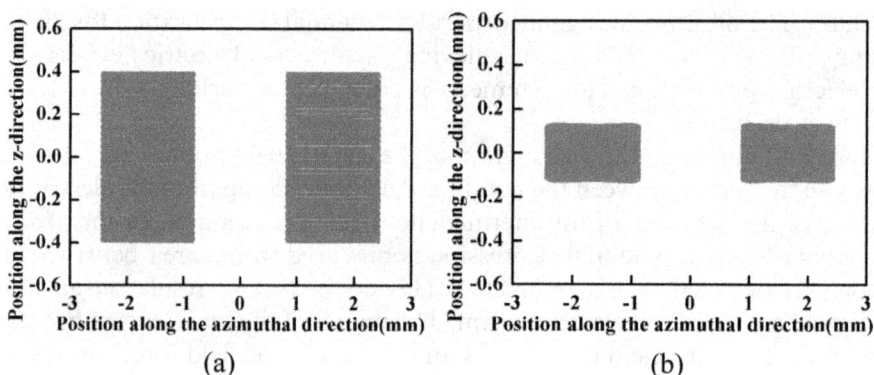

Figure 3.44. Cross-sections of the electron beam: (a) cathode emission surface; (b) beam waist position ($r = 95$ mm). Reproduced from [13]. CC BY 4.0.

respect to the electron beam center, and the electric fields experienced by the two radial electron beams are also completely symmetric, as depicted in figure 3.42(b). By adjusting the middle focusing pole to compensate for the electric field in the electron gun region, the tracks of the two sets of electron beams can be prevented from deflecting toward the center, as illustrated in figure 3.43(b).

Figure 3.44 presents the 2D particle distribution of the radial twin electron beam, including the distribution on the φ–z plane at the cathode emission surface and the beam waist position ($r = 95$ mm). At the beam waist, the electron beam size is found to measure 1.6 mm × 0.3 mm, aligning with the ideal electron beam size in the beam interaction calculation model and meeting the design criteria. The compression ratio of the electron beam in the z-direction is approximately 2.5:1, with a relatively small compression ratio in the φ-direction. This dual-direction compression ensures that the electron beam's waist occurs at nearly the same radial position in both directions, simplifying the design of the subsequent focusing system. The total emission current of the radial double electron beam converges after 30 iterations, reaching a final emission current of 0.395 A, which only differs by 1.25% from the working current of 0.4 A calculated by the beam–wave interaction model.

3.5.1.2 Flow of double angular radial electron beams

The focusing magnetic field is incorporated into the calculation model of the radial double-beam electron gun to analyze the flow state of the radial double electron beam. Notably, the radial dual-beam focusing discussed in this section is conducted without any signal input to the SWS. As depicted in figure 3.45, the trajectory of radial twin electron beams with radial PCM focusing reveals that the two radial electron beams maintain relative independence during the transmission process, with no mutual interference in angular divergence. Figure 3.46 illustrates the envelope of the radial double electron beam in the φ- and z-directions in the presence of the radial PCM field. It is evident that for an electron beam channel with a length of 13 mm, no electrons reach the boundary of the electron beam channel. This signifies

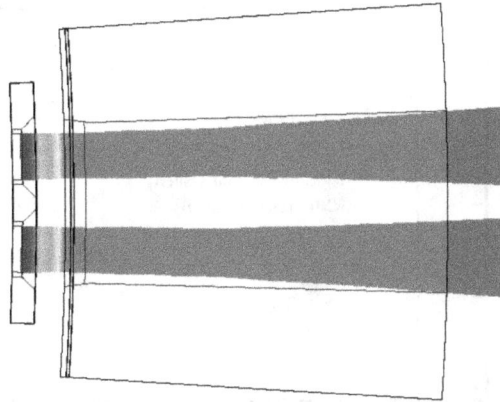

Figure 3.45. Radial twin electron beam subjected to radial PCM focusing.

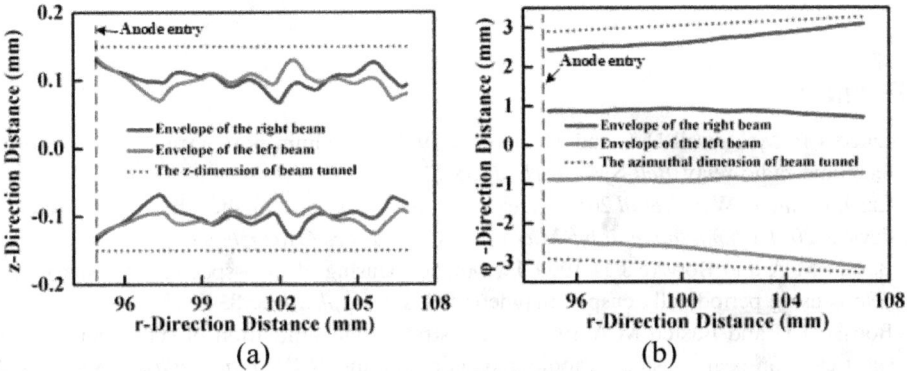

Figure 3.46. Envelopes of a radial double electron beam in the presence of a focused radial PCM field. (a) r–φ plane and (b) r–z-plane. Reproduced from [13]. CC BY 4.0.

that the radial double electron beam achieves a flow rate of 100% under a radial PCM field.

In practical TWT electron guns, the electron beam is emitted with a certain initial velocity due to the hot cathode. The initial thermal velocity of the electrons can impact the focusing of the electron beam. Therefore, the Maxwell–Boltzmann thermal velocity distribution is introduced into the calculation model in this section. Following the conversion relationship between voltage and cathode temperature given in the research literature, the cathode temperature is set to 1400 K. Figure 3.47 illustrates the influence of the thermal velocity on the envelope of the beam in the r–z plane. When the electron thermal velocity distribution is included, the girth of the envelope is slightly larger than in the absence of this effect. However, given the very small increase in girth, the impact of the thermal velocity of the electrons can be deemed negligible in the calculation model.

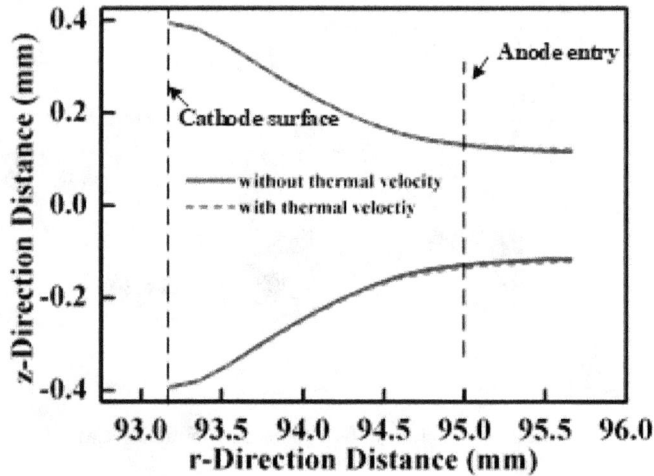

Figure 3.47. Influence of the thermal muzzle velocity on electron beam focus. Reproduced from [13]. CC BY 4.0.

References

[1] Laico J P, McDowell H L and Moste C R 1956 A medium power traveling-wave tube for 6,000-mc radio relay *Bell Syst. Tech. J.* **35** 1285–346

[2] Lai J, Gong Y, Wei Y *et al* 2011 An electron optical system for sheet beam vacuum electron devices *2011 China-Japan Joint Microwave Conf. Proc. (Hangzhou)* pp 1–3

[3] Basten M A and Booske J H 1999 Two-plane focusing of high-space-charge sheet electron beams using periodically cusped magnetic fields *J. Appl. Phys.* **85** 6313–22

[4] Booske J H and Basten M A 1999 Demonstration via simulation of stable confinement of sheet electron beams using periodic magnetic focusing *IEEE Trans. Plasma Sci.* **27** 134–5

[5] Humphries S, Russell S, Carlsten B and Earley L 2005 Focusing of high-perveance planar electron beams in a miniature wiggler magnet array *IEEE Trans. Plasma Sci.* **33** 882–91

[6] Zhao D 2010 Research on feasibility of closed and offset PCM focusing structures for sheet electron beams *Acta Phys. Sin.* **59** 1712–20

[7] Langmuir I and Blodgett K B 1923 Currents limited by space charge between coaxial cylinders *J. Phys. Rev.* **22** 347356

[8] Chen X *et al* 2004 Approximate analytical solutions for the space-charge-limited current in one-dimensional and two-dimensional cylindrical diodes *Phys. Plasmas* **11** 3278–82

[9] Wang S, Gong Y, Wei Y and Duan Z 2012 Study on the radial-sheet-beam electron optical system *IEEE Trans. Plasma Sci.* **40** 3442–8

[10] Wang S *et al* 2014 Study of low-voltage radial convergent sheet electron optical system *IEEE Trans. Plasma Sci.* **42** 1847–53

[11] Li X *et al* 2015 Design of the radial divergent sheet beam electron optical system with cylindrical emission surface *2015 40th Int. Conf. on Infrared, Millimeter, and Terahertz waves (IRMMW-THz) (Hong Kong, China)* pp 1–2

[12] Li X *et al* 2017 Study on radial sheet beam electron optical system for miniature low-voltage traveling-wave tube *IEEE Trans. Electron Devices* **64** 3405–12

[13] He T *et al* 2021 Electron-optical system for dual radial sheet beams for Ka-band cascaded angular log-periodic strip-line traveling wave tube *AIP Adv.* **11** 035325
[14] Gilmour A S 1994 Principes of Traveling Wave Tubes (Artech House)
[15] Wang H *et al* 2023 Beam-wave interaction and electron optical system investigation for a W-band planar TWT *24th Int. Vacuum Electronics Conf. (IVEC) (Chengdu, China)* pp 1–2

Chapter 4

Fabrication technologies

Typically, a traveling-wave tube (TWT) consists of more than a hundred components, and the manufacture of one component may require several different processes to be used. Manufacturing and assembly tolerances play a crucial role for planar TWTs operating in bands above the millimeter wave band, and small tolerances can lead to significant differences in design performance. Planar TWTs have attracted much attention because of their simple structure and potential for mass production [1, 2]. Various advanced manufacturing techniques have been tried to manufacture them [3–6]. This chapter presents attempts to fabricate the planar SWSs mentioned in chapter 2.

4.1 Suspended double-microstrip meander line

The suspended double-microstrip meander-line slow-wave structure (SWS) consists of a microstrip meander line and a metal enclosure, which have to be fabricated separately. To fit the microstrip into the enclosure, the latter should be machined in two halves, as shown in figure 4.1.

The process of preparing the suspended double-microstrip meander-line SWS on a ceramic substrate is as follows:

(1) Clean an alumina ceramic sheet that has a diameter of 20 mm and a thickness of 0.2 mm, as shown in figure 4.2(a).
(2) Perform double-sided magnetron sputtering to add a coating to the alumina ceramic sheet; the coating consists of a 500 nm titanium layer and a 1 μm copper layer, as shown in figure 4.2(b).
(3) Electroplate the coated alumina ceramic with a 10 μm thick copper layer on the surface, as shown in figure 4.2(c).
(4) Laser ablate the copper layer on the surface to form a symmetrical meander microstrip line.
(5) Cut the alumina ceramic into the corresponding shape.

doi:10.1088/978-0-7503-5452-3ch4

(a)

(b)

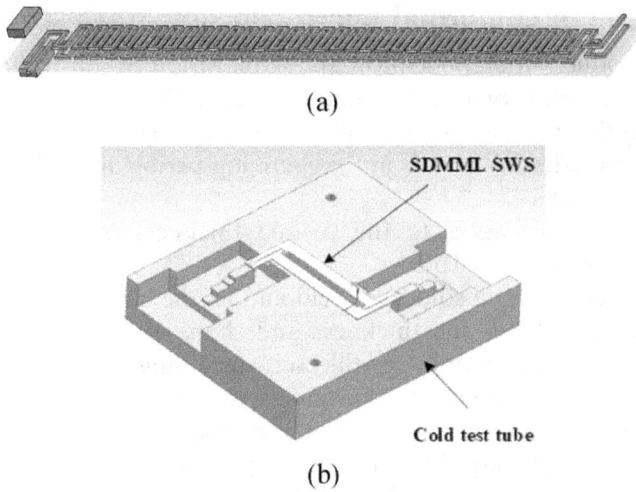

Figure 4.1. (a) Simulation model and (b) fabrication model of a suspended double-microstrip meander-line SWS.

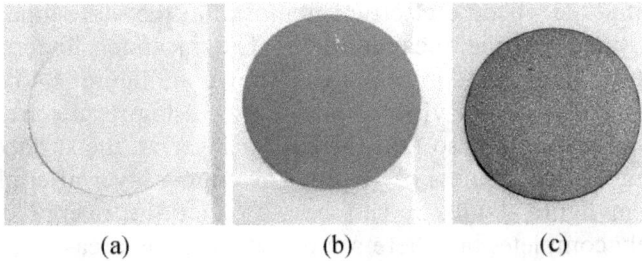

(a)   (b)   (c)

Figure 4.2. Alumina ceramic sheet: (a) after cleaning; (b) after magnetron sputtering is used to add a titanium–copper layer; (c) after electroplating is used to add a copper layer.

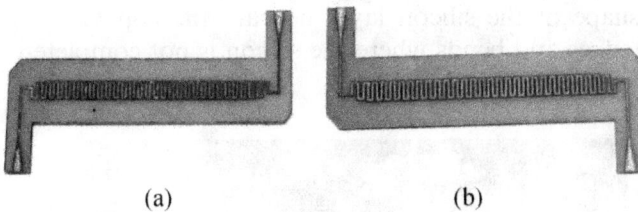

(a)   (b)

Figure 4.3. Suspended double-microstrip meander-line SWS. (a) Front layer; (b) back layer.

Figure 4.3 shows the suspended double-microstrip meander-line SWS. The front and back structures of the meander-line SWS are respectively shown in figure 4.3(a) and (b). It can be seen that the laser processing is consistent and that the two layers of the microstrip meander line are completely symmetric.

4.2 Fabrication of conformal microstrip slow-wave structures

This section introduces the application and exploration of ion-beam etching (IBE) technology in the fabrication of conformal microstrip SWSs ('conformal' means that the substrate has the same shape as the meander line). The manufacturing process of a conformal microstrip SWS with an angular log-period meander line using IBE technology is as follows:

(1) Thin the silicon wafer to the desired thickness of 80 μm, which is the controllable depth of IBE.

(2) Use low-temperature solder to weld and fix the silicon sheet onto the metal support plate. Since the thickness and shape of the conformal microstrip structure cannot guarantee the self-sustaining property of the whole structure, it is necessary to continue welding and fixing the silicon wafer and metal support to ensure that the whole structure does not become deformed and damaged in subsequent processing.

(3) A copper layer of 1 μm thick is deposited on the surface of the silicon wafer by plasma-enhanced chemical vapor deposition (PECVD), as shown in figure 4.4(a). The photolithographic mask is designed and placed on the surface of the silicon wafer, and the whole structure is treated using the IBE process. The copper layer on the surface is etched into the corresponding shape, and the silicon dielectric is removed at the same time, except for the part covered by the meander line. The meander line of a conformal microstrip after IBE processing is shown in figure 4.4(b), wherein the meander line is clearly visible and the processing results are good.

(4) Electroplate the SWS so that the copper layer on the surface of the silicon layer reaches a thickness of 10 μm. The copper layer after electroplating is shown in figure 4.4(c). It can be seen that the overall copper layer is relatively complete, but there are defects in some areas.

Figure 4.5 shows a conformal microstrip angular log-period meander-line SWS after all processing has been completed. It can be seen that the shape and morphology of the electroplated copper layer on the surface meet the design requirements. However, the shape of the silicon layer beneath the copper layer is flawed, and there are some edges and bends where the silicon is not completely etched away.

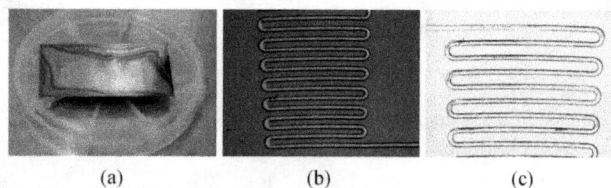

(a)   (b)   (c)

Figure 4.4. Photo of an SWS during processing. (a) Silicon wafer after the completion of PECVD; (b) conformal microstrip meander line after the completion of IBE; (c) microstrip meander line after electroplating [12], reproduced courtesy of The Electromagnetics Academy.

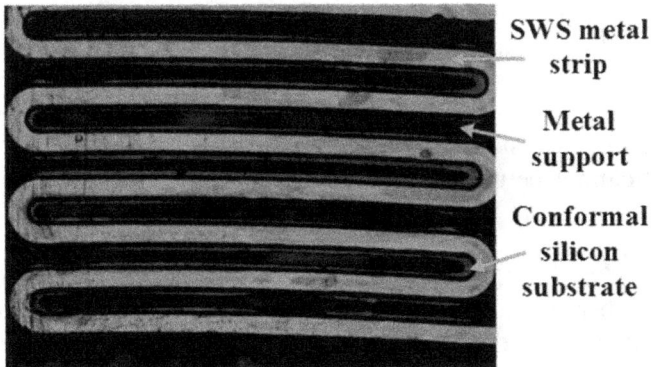

Figure 4.5. Sample of a conformal microstrip angular log-period meander-line SWS [7], reproduced courtesy of The Electromagnetics Academy.

Figure 4.6. Residual silver solder on the surface of the metal support sheet.

A photo of the metal support sheet is shown in figure 4.6. It can be seen that there is still a lot of solder residue on the welding surface that faces the silicon sheet. This solder is present on the surface of the metal support sheet and cannot be removed. An aggressive removal process would also affect the welding condition of the conformal microstrip SWS and destroy the connection between the SWS and the metal support; therefore, solder residue will be present on the metal support sheet, and a corresponding improvement should be made as part of the subsequent processing.

There are still many problems with conformal microstrip angular log-period meander-line SWSs fabricated by IBE technology, which cannot be used in TWTs. First, there is the problem of the silicon dielectric. While exploring the process, due to the high price of monocrystalline silicon substrates and the difficulty in obtaining them, a conventional silicon substrate was used in the experiment. This substrate, which is a kind of semiconductor material, has an adverse effect on the transmission of electromagnetic waves, and no electromagnetic wave transmission was detected in the transmission test. The second is the problem of welding the silicon wafer to the metal support sheet. The silver solder used at present is a kind of low-temperature solder that cannot withstand the welding exhaust and other high-temperature processes, so it cannot be used in tests of TWT loading.

4.3 Angular log-period meander strip-line SWS

The angular log-period meander strip-line SWS consists of the following components: a metal log-period strip, two dielectric support rods, and a metal enclosure with a pair of ridged waveguide input/output coupling structures and a slot for the SWS [8, 9]. The material properties of each part have significantly different characteristics, so we use separate processing and assembly processes.

The main fabrication process includes the following steps: (1) the preparation of the log-period meander strip line; (2) local metallization and cutting/shaping of the dielectric support rod; and (3) machining of the metal enclosure. The more difficult steps are the preparation of the log-period meander strip line and the dielectric support rod, and the welding and assembly of the two parts.

According to the technological requirements of vacuum electronic devices, multichannel high-temperature welding and high-temperature fume exhaust are required for the angular log-period meander strip line. The material selection of the strip line affects the stability, reliability, and operational performance of the whole SWS. After comprehensive consideration, a molybdenum–rhenium alloy with high hardness, a high melting point, and no deformation was selected as a raw material. Because of the small size and flat structure of the angular log-period meander strip line, some common processing methods used with SWSs, such as helix, were unsuitable for the strip line. In view of the planar property of the angular log-period meander strip-line SWS, a processing scheme was developed based on laser processing of the strip line. The laser cutting method precisely machined the angular log-period meander strip-line SWS and effectively controlled processing errors.

The preparation process used for the angular logarithmic ribbon is as follows:

(1) Pre-treatment: clean and flatten the 0.1 mm thick molybdenum–rhenium alloy sheet to ensure the smoothness of the material.

(2) Laser processing: cut and ablate the 0.1 mm thick alloy sheet to produce the shape of the logarithmic strip lines using a picosecond laser, as shown in figure 4.7(a).

(3) Peeling treatment: immerse the cut alloy sheet and the complete strip line into an anhydrous ethanol solution, then clean and separate the SWSs using an ultrasonic cleaning machine.

(4) Deep cleaning: there are many burrs and bumps on the surface and edge of the laser-cut strip line, which would seriously affect the transmission characteristics of the SWS. In order to eliminate transmission loss caused by burrs and bulges, it is necessary to acid wash the SWS, as shown in figure 4.7(b).

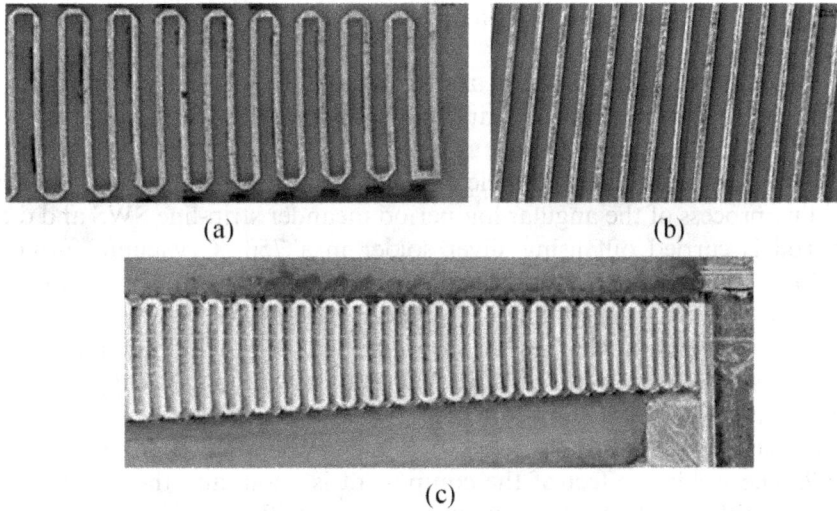

Figure 4.7. Angular log-period meander strip-line SWS during fabrication. (a) After laser cutting; (b) after deep cleaning treatment; (c) after surface sputtering treatment.

(5) Surface sputtering treatment: magnetron sputtering is used to form a 10–20 μm copper layer on the cleaned angular logarithmic strip-line surface to improve the transmission characteristics of the SWS and reduce metal loss, as shown in figure 4.7(c).

In order to support the angular log-period meander strip-line SWS, the dielectric support rod needs to be welded tightly to the strip line. The structural characteristics of the strip-line SWS dictate that the contact mode between the strip line and the dielectric support rod is a multipoint discrete contact, and the contact area of each point is very small. Therefore, the preparation of the dielectric support rod determines the welding quality of the strip line and the dielectric support rod to a certain extent. After many machining and welding experiments, a preparation technology was selected for the dielectric support rod. Due to the small size of the angular log-period meander strip-line SWS and its easy deformation, conventional solder coating, spot welding, and other processes are unsuitable for welding the support rod to the metal meander line. In order to weld the dielectric support rod to the metal slow-wave line, a ceramic metallization process was developed to realize polymetallic solder points on the dielectric support rod.

The preparation process for the dielectric support rod is as follows:

(1) Coat a 2–4 μm molybdenum–manganese layer onto an alumina ceramic plate 20 mm in diameter using screen-printing technology to form a metal base.

(2) Form a silver layer with a thickness of 5–8 μm on the surface of the alumina wafer using an electroplating treatment; this functions as the solder layer.

(3) Cauterize the silver solder layer by laser to form some discrete flaky solder spots.

(4) Use a laser to cut the alumina disc. Cut out a dielectric support rod 2 mm thick.

(5) Clean the dielectric support rod to remove the oxide layer formed on the surface of the metal solder during processing.

Figure 4.8 shows the finished dielectric support rod with the silver solder joint. The welding process of the angular log-period meander strip-line SWS and dielectric support rod is carried out using silver solder in a 750 °C vacuum furnace. The angular log-period meander strip-line SWS is welded to the upper ridge waveguide of the tube shell using a silver–copper solder sheet and a spot welding machine. The upper and lower parts of the tube shell, input and output windows, and flanges are welded. The solder groove of the tube shell is filled with silver–copper solder wire and welded in a 750 °C vacuum furnace.

The angular log-period meander strip-line SWS after welding is shown in figure 4.9. The welding effect of the component is good, and the welding strength of the metal solder joint and the slow-wave line meets the requirements. However, it can be seen that there is some slight deformation in the transition section of the SWS, which needs to be detected in the subsequent transmission characteristics test.

The assembly of the angular log-period meander strip-line SWS and the metal enclosure is now complete. See figure 4.10 for enlarged assembly photos of the electron beam tunnel part and the impedance transformation part. Figure 4.11 shows a slow-wave system in which all components are assembled and welded.

After the welding and assembly of the angular log-period meander strip-line SWS, the transmission performance of the SWS was tested experimentally, and the experimental results were compared with the simulation results to ensure that the transmission performance of the SWS met the design requirements and to guarantee the reliability and authenticity of the subsequent whole-tube experiments.

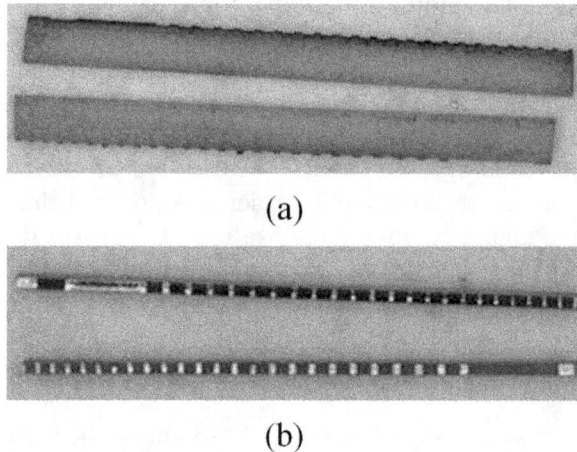

(a)

(b)

Figure 4.8. Dielectric support rod with a silver solder joint. (a) Front side of the dielectric rods; (b) dielectric rod solder surfaces. Reprinted from [8], Copyright (2020), with permission from Springer.

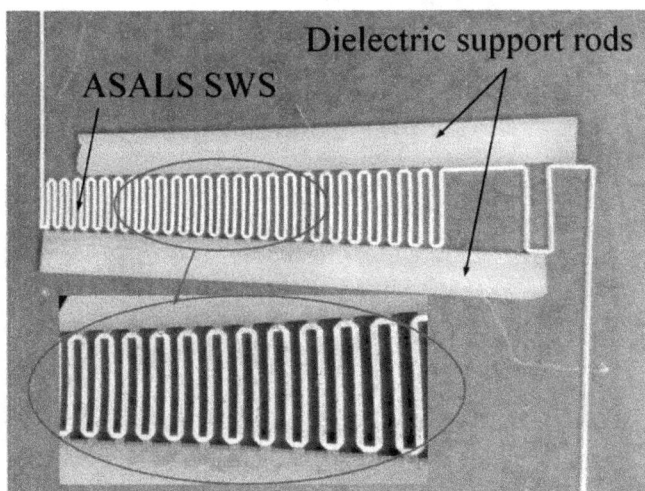

Figure 4.9. Angular log-period meander strip-line SWS after welding. Reprinted from [8], Copyright (2020), with permission from Springer.

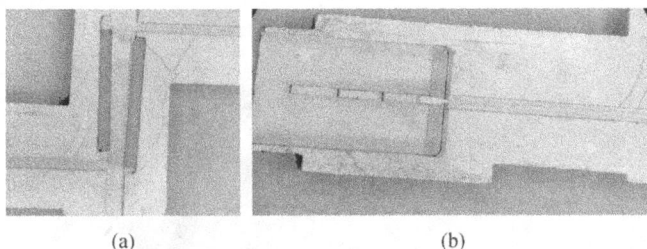

(a)   (b)

Figure 4.10. Angular log-period meander strip-line SWS. (a) Access area; (b) connection area between the strip line and the ridge waveguide. Reprinted from [8], Copyright (2020), with permission from Springer.

Following assembly and welding, the SWS was connected to the test system. The test results are shown in figures 4.12 and 4.13. They are the voltage standing-wave ratio (VSWR) of the input and output ports and the transmission loss of the slow-wave system, respectively. In the full Ka band (26.5–40 GHz), the port VSWR is generally less than 2, and the transmission loss is less than 2.5 dB.

Figure 4.14 shows a comparison between the experimental results and the simulated results for the transmission characteristics of the slow-wave system. Figure 4.14(a) shows a comparison between the experimental and simulated transmission coefficients. The experimental result is about 0.5 dB larger than the simulation result at the low-frequency end, and the difference between the two is small at the high-frequency end. Both errors are within the allowable range and do not affect the experimental results for the thermal measurement of the whole tube. Figure 4.14(b) shows a comparison between the experimental reflection coefficient and the simulated result. In the frequency range of 26–40 GHz, the experimental reflection coefficient is consistent with the simulated result, but the experimental

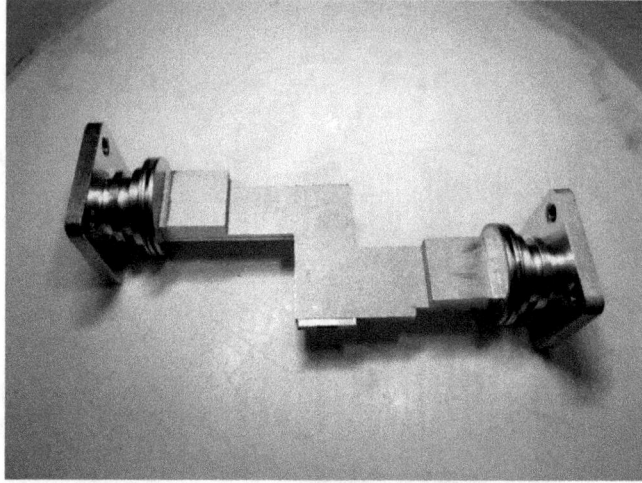

Figure 4.11. Photo of a slow-wave system after assembly and welding. Reprinted from [8], Copyright (2020), with permission from Springer.

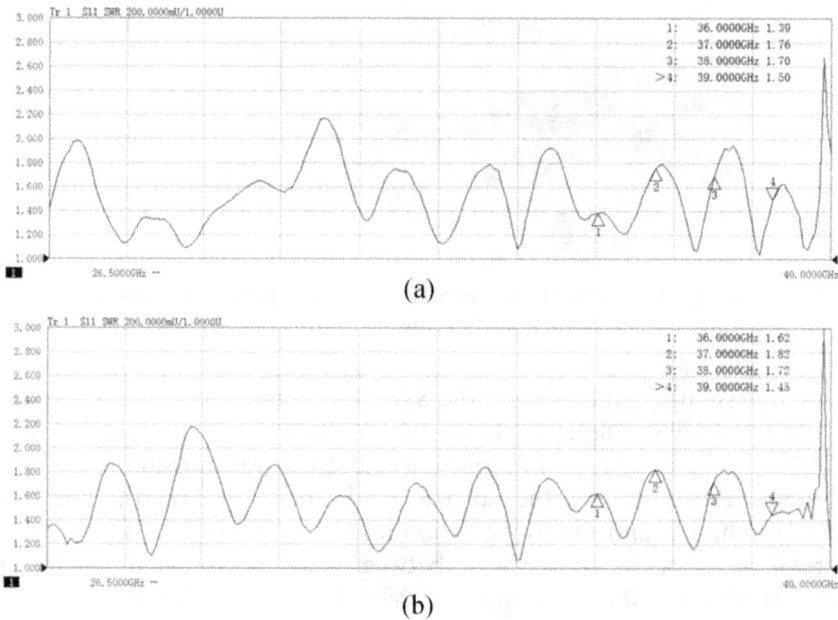

(a)

(b)

Figure 4.12. Voltage standing-wave ratio of the slow-wave system. (a) Input port; (b) output port.

result is generally lower than −10 dB, and the simulated result is lower than −15 dB, leading to a difference of −5 dB.

The main reasons for the difference between the experimental results and the simulation results are: (1) the size deviations, surface roughness, and burrs generated by machining have a negative effect on the transmission characteristics and increase

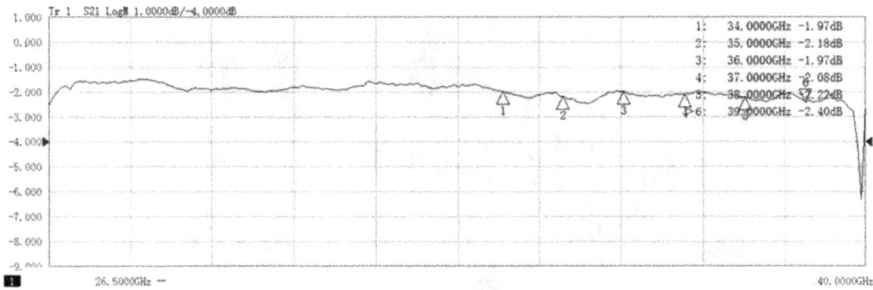

Figure 4.13. Transmission loss of the angular log-period meander strip-line SWS.

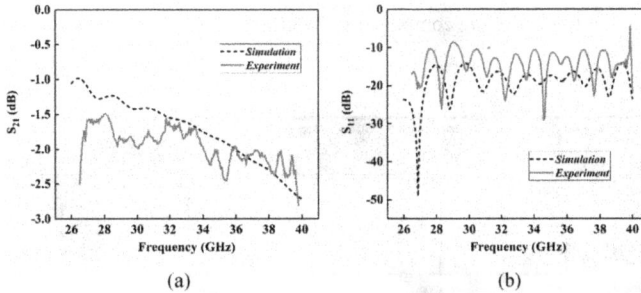

Figure 4.14. Comparison between the simulated and experimental results for the transmission characteristics of the angular log-period meander strip-line SWS. (a) Transmission coefficient; (b) reflection coefficient. Reprinted from [8], Copyright (2020), with permission from Springer.

reflection and loss. (2) In order to increase welding reliability (and thus reduce the probability of air leakage), the shell is plated with nickel, but the nickel layer increases the attenuation of high-frequency microwave signals and increases the overall insertion loss. (3) The slow-wave line is still slightly deformed by the welding process, and this structural deformation increases the discontinuity of the overall structure, resulting in increased reflection.

4.4 Quartz-based microstrip angular log-period meander-line SWS

Based on the structural parameters determined by a simulated design, the physical components of an angular log-period meander-line SWS with a 0.254 mm thick quartz substrate were designed and prepared for cold measurement, as shown in figure 4.15.

The angular log-period meander-line SWS was placed in the cavity of the test bottom plate. The dimensional height of the cavity was the sum of the height of the air cavity (h) and the thickness of the quartz bottom plate (d). A coaxial window was fixed at the position corresponding to the end of the tapered microstrip line through a hole at the corresponding position of the test bottom plate. The microstrip line was connected to the inner conductor of the coaxial window by a gold wire bonding process using 0.025 mm thick gold wire. The test cover plate was a metal plate provided with a hole that was tightened through a threaded hole on the test bottom plate.

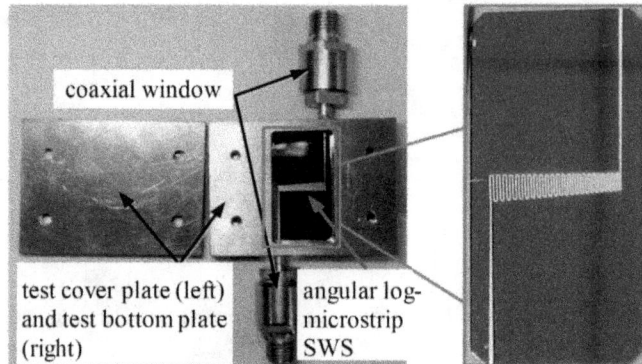

Figure 4.15. Photo of the test assembly components. Reprinted from [9], Copyright (2018), with permission from IET.

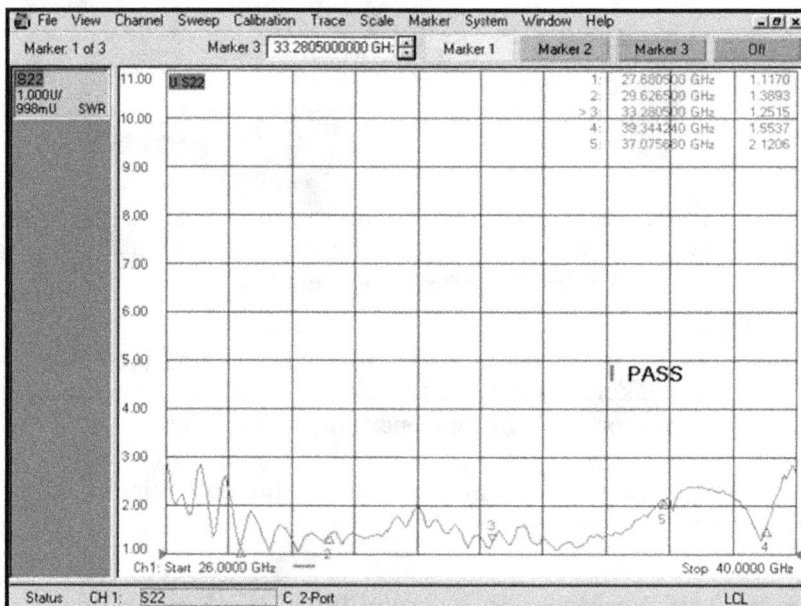

Figure 4.16. VSWR screenshot of the input port. Reprinted from [9], Copyright (2018), with permission from IET.

The VSWR test results for this sample are shown in figure 4.16. It can be seen from the data curve in the figure that the VSWR value is less than 3.0 in the full frequency band and less than 2.0 in the range of 28.1–35.8 GHz. The difference between the test results and the simulated results is mainly due to the uncertainty caused by the gold wire bonding and the position of the gold wire which was manually welded in the fabrication process.

After the quartz-based angular log-period meander-line SWS was assembled in the tube body, a hydrogen furnace welding process was required for the whole tube assembly process. In order to verify the adhesion degree of the metal meander lines on the quartz surface, the SWS was placed in a hydrogen furnace for a high-temperature test. See figure 4.17 for a photo taken in the furnace at a high temperature. The reference solder was silver–copper 28 (AgCu28), and the maximum temperature of the hydrogen furnace was set to 780 °C. Because the melting point of quartz is 1750 °C, quartz does not exhibit softening, deformation, or other conditions at the test temperature. When the hydrogen furnace reached 780 °C, the state of the reference solder was observed to change from filamentous to molten; the solder was then allowed to cool to room temperature naturally.

Figure 4.18 is a partial photo of the sample taken in hydrogen furnace after the high-temperature test. The photo shows that the metal meander-line did not suffer from peeling, loosening, or detachment after passing through the high-temperature environment (the melting-point temperature of AgCu28 is 780 °C). However, a circular defect appeared on the quartz substrate near the input end, as shown in figure 4.19.

A partial magnification of the circular defect on the quartz substrate shows that there is a melting-point hole in the center, which has penetrated the quartz substrate.

It is stated in the literature related to the oxygen-containing compounds of silicon that silica does not react to chlorine (Cl_2), hydrogen (H_2), acids, or most metals at room temperature and slightly higher temperatures. Among these reactions, the

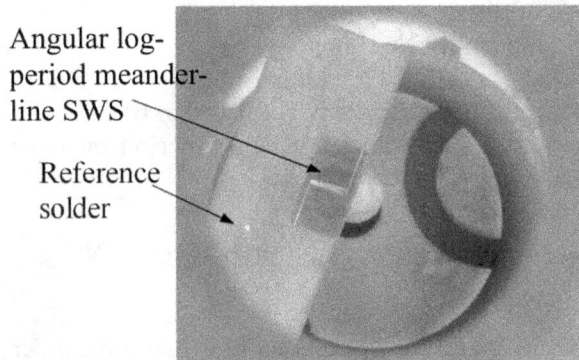

Figure 4.17. Quartz-based angular log-period meander-line SWS in a high-temperature test (hydrogen furnace).

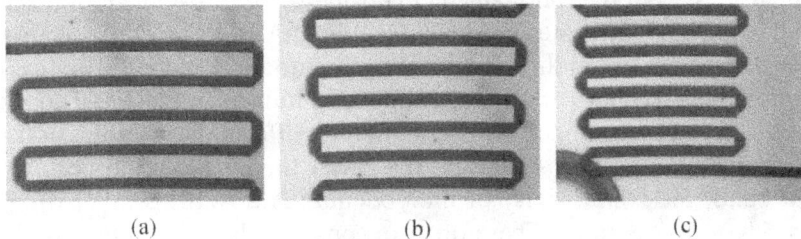

Figure 4.18. Photos of samples after the high-temperature test. (a) Output terminal; (b) intermediate segment; (c) input end.

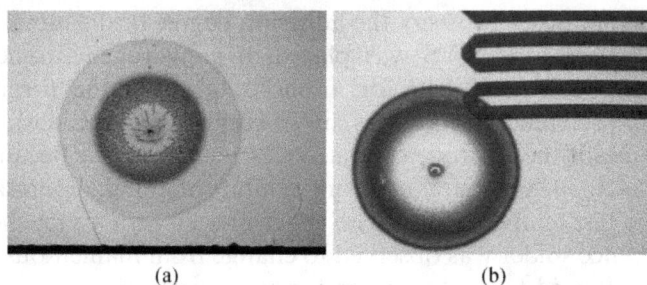

(a)   (b)

Figure 4.19. Defect in the quartz sheet. (a) Reverse surface to that of the microstrip line; (b) microstrip line surface.

conditions for the reaction with hydrogen are not explicitly given. In 2017, a patent for the preparation of nano silica vapor mentioned that at a reaction temperature of 1000 °C–1800 °C, chlorosilane ($SiHCl_3$) was hydrolyzed with water vapor (H_2O) to obtain silica vapor particles.

It can be inferred from the above description that the formation of the weld hole was not caused by a reduction reaction between the local quartz substrate and hydrogen in the high-temperature hydrogen atmosphere. The probable cause was a defect in the quartz substrate, which melted at high temperature or reacted with hydrogen. The molten or reaction products then diffused in the quartz substrate to form radiating rings of different colors.

The revelation from this phenomenon is that a high-temperature screening process for the quartz substrate should be added to the future process of tube production so that defective unqualified products can be discovered in advance and the subsequent process of preparing angular log-period meander-line SWSs using those substrates can be avoided.

4.5 Attenuator-supported meander strip-line SWS

4.5.1 Fabrication process

The meander strip-line SWS mainly consists of a metal meander strip line and an attenuator [10]. The attenuator can be prepared by the physical deposition of a carbon layer on a block of the selected dielectric, which is relatively simple. The production of strip lines is more complex and critical. Its main material selection criteria are discussed in the following. (1) Hardness. The strip line requires a certain degree of flexibility to facilitate subsequent welding. At the same time, it also needs a certain hardness and the ability to maintain a smooth structure after processing. Molybdenum, which is hard and tough, appears suitable. (2) Electrical conductivity. Because the conductivity of the strip line directly affects the transmission loss, the higher the conductivity of the material, the better the transmission performance is. In the Ka band, the conductivity of molybdenum is 2×10^7 S m^{-1}, while that of copper is 5.8×10^7 S m^{-1}. Therefore, to consider the advantages of both, the structure proposed as the strip-line material in this chapter is a magnetron-sputtered copper layer on the surface of a molybdenum wire.

The complete process used to produce the strip line is as follows:

(1) Select a molybdenum sheet with the required thickness. Press it flat with a heavy weight, and then sinter it at a high temperature to ensure the flatness of the molybdenum sheet.

(2) Cut molybdenum sheet with a picosecond laser to make a meander molybdenum wire with the required design pattern.

(3) Magnetron sputter a copper layer that coats the entire surface of the molybdenum wire.

The simple process of laser cutting is used. The structural pattern is sent to the cutting system, and a high-power pulsed laser spot is focused on the edge of the structure such that the spot diameter is about 12–14 μm. At this point, the edge of the sample is vaporized by continuous laser ablation, and eventually a hole the size of the light spot is formed. After completing the ablation of a spot, the beam travels along a programmed path, ablating the edge of the sample. Finally, point-to-point lines are connected to cut the desired pattern from the substrate. In figure 4.20, the material substrate is shown in blue, the required pattern is shown in purple, and the light spot is red. When the movement steps of the spot are densely clustered, the edge of the cut sample is smooth, but if the steps are too dense, it is easy to overburn and oxidize the surface of the material. If this occurs, it is necessary to add an inert protective gas. If the movement steps are sparse, the cut will be rough, and it may not even be possible to remove the sample from the substrate. Therefore, the process needs to adjust the optimal process parameters according to the substrate material and thickness, repeating tests of the laser power and stepping speed. Some key parameters obtained after optimization are as follows: a laser power of 15.2 W, a frequency of 40 kHz, a cutting speed of 300 mm/s, and a spot focus movement of 4 μm after 40 repetitions of single point processing.

The basic principle of magnetron sputtering coating is that the sample substrate is used as the anode, the target is connected to a negative high voltage and functions as the cathode, and the argon gas in a vacuum furnace is ionized by electrons from the target material to produce positive argon ions. Because the

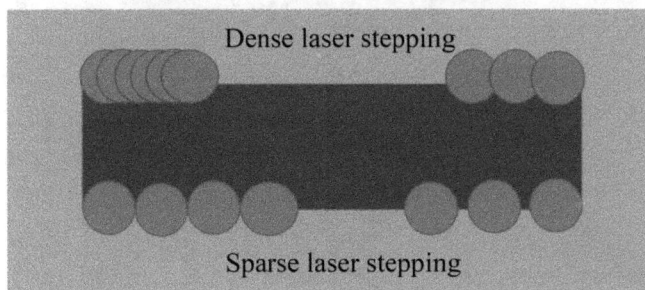

Figure 4.20. Schematic diagram showing how the cut strip line varies with the degree of asynchronous laser spot progression.

target is subjected to a negative voltage, high-energy argon ions bombard the target; their kinetic energy knocks out atoms from the surface of the target. Because the sputtered target atoms have high energy when deposited on the sample substrate, high-density deposited films can be formed with high adhesion. By controlling the sputtering power, the coating rate can be adjusted. Finally, the parameters used for the strip line coating of the molybdenum material were as follows: a DC power of 420 W, a constant voltage of 400 V, a background temperature of 150 °C–200 °C, a coating rate of 0.75 μm h^{-1}, and a speed of 25 rpm. The resulting copper layer was about 5 μm thick.

The overall assembly process of the SWS was as follows: a slot is cut at the bottom of the metal shell to accept the dielectric substrate or attenuator, and the strip line is placed above the attenuator by positioning the clamp so that the end of the strip line is level with the first ridge inside the rectangular waveguide. In order to weld the ridges of strip lines and waveguides, platinum solder sheets are generally used as the intermediate layer for spot welding. A high-current spot welder is then used to complete the welding. Finally, the upper part of the slow-wave shell is covered and fastened with a splint to complete the production of the SWS. An assembly diagram of the SWS is shown in figure 4.21.

In order to study the performance and strip-line loss of the prepared attenuator, two SWSs were prepared, one of which was supported by an attenuator and the other of which was supported by a pure dielectric substrate (see figures 4.22(a) and (b), respectively). Figure 4.22(b) also shows the attenuator details and the contact details between the end of the strip line and the platinum solder sheet.

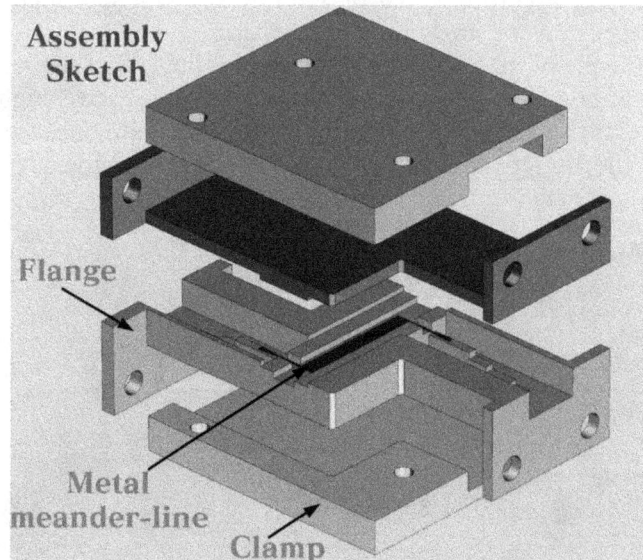

Figure 4.21. Overall assembly diagram of the SWS. Reproduced from [10]. CC BY 4.0.

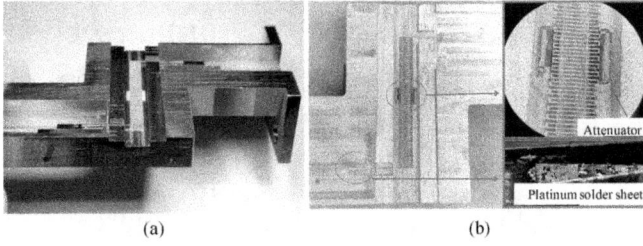

(a)   (b)

Figure 4.22. Manufacture of planar SWSs supported by two different materials. (a) Pure dielectric substrate support; (b) attenuator support. Reproduced from [10]. CC BY 4.0.

4.5.2 Measurement and analysis of surface roughness

Since the conductivity of the material directly determines the transmission loss of the SWS, this section describes the quantitative characterization of the surface conductivity of the strip line, which is performed by testing the surface roughness of the material. In addition, the processing of the prepared strip line is qualitatively observed. The equivalent conductivity σ_{ef} is given by equation (4.1), where σ is the inherent conductivity of the material, R_S is the roughness of the material surface, δ is the skin depth of the conductor as calculated using equation (4.2), μ is the permeability, and ω is the angular frequency:

$$\sigma_{ef} = \frac{\sigma}{\left(1 + \frac{2}{\pi}\arctan\left(1.4 \times \left(\frac{R_S}{\delta}\right)^2\right)\right)^2} \tag{4.1}$$

$$\delta = \sqrt{\frac{2}{\omega\mu\sigma}}. \tag{4.2}$$

Therefore, in order to improve the equivalent conductivity of the material, it is necessary to increase the intrinsic conductivity of the material, σ, or reduce the surface roughness, R_S. The original pure molybdenum material is replaced by a magnetron-sputtered copper layer on the surface of molybdenum for the prepared strip line material, which can improve the equivalent conductivity of the material. This is related to the relationship between the skin depth of the material and the thickness of the sputtered copper layer. It is calculated that the skin depth of copper should be 340 nm for the Ka band. Therefore, the sputtered copper layer only needs to be thicker than 340 nm to ensure that the active area of the electromagnetic wave action consists of copper. In the actual preparation process, the thickness of the sputtered copper layer is about 5 μm, so it is sufficient to meet this condition.

We now turn to R_S (surface roughness). The original molybdenum sheet is made by pressing molybdenum powder, and the amount of grain in the upper and lower surfaces is relatively high. At the same time, the molybdenum strip is prepared by cutting the molybdenum sheet on the side by laser (as shown in figure 4.23(a)), so there are a series of burn marks on the side, making the side rougher than the upper and lower surfaces. By sputtering a copper layer, the upper and lower surfaces and

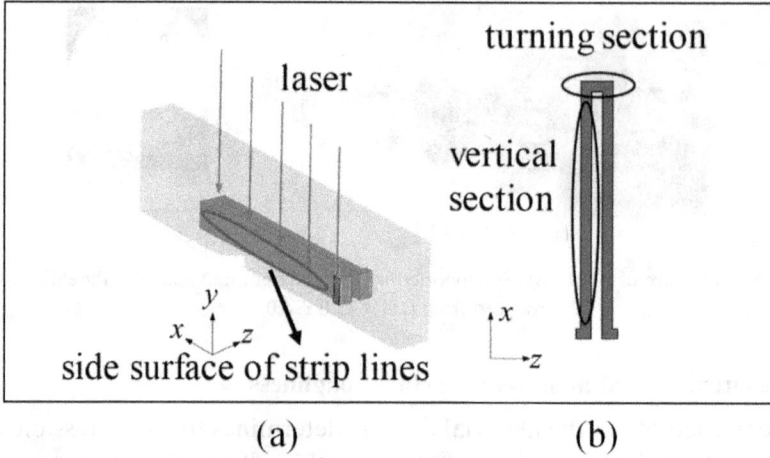

Figure 4.23. Principle of laser cutting and a structural diagram. (a) Schematic diagram of the laser cutting process used to produce strip lines; (b) different labeled parts of the ribbon line. Reproduced from [10]. CC BY 4.0.

side areas can be evenly covered, which effectively improves the surface roughness. Therefore, the roughnesses of a pure molybdenum wire and a molybdenum wire with a sputtered copper layer are studied.

The surface roughness R_S is usually represented by the surface average roughness S_a in the test area, which is determined using formula (4.3):

$$S_a = \left(\frac{1}{NM}\right)\sum_{i=1}^{N}\sum_{j=1}^{M}|Z_{ij}| \tag{4.3}$$

where N and M are the numbers of sampling points in the two directions perpendicular to each other in the plane of the test area, respectively, and Z is the height from the test point to the datum plane. The surface average roughness S_a can be calculated by summing and averaging. In the testing described in this section, an Olympus LEXT 5000 laser confocal microscope was used to test the roughness of the strip lines of pure molybdenum material and the roughness of the magnetron-sputtered copper layer on the surface of the molybdenum wire; the upper surface and the side surface were used, respectively. The upper surface was divided into a turning section and a vertical section, as shown in figure 4.23(b). The surface roughness of the strip line was obtained by means of multipoint sampling and averaging. The test results are discussed below.

Figures 4.24 and 4.25 respectively show the pure molybdenum strip line and the magnetron-sputtered copper on the surface of the molybdenum wire. Both figures show the roughness of a turning section and a vertical section of the upper surface. Figures 4.26 and 4.27 respectively show the roughness of the strip lines of these two materials in three sampling areas on the side. All sampling and measurement results are summarized in table 4.1.

It can be concluded from table 4.1 that: (1) the roughness of the upper surface is indeed better than that of the side, which is an inevitable defect caused by the use of

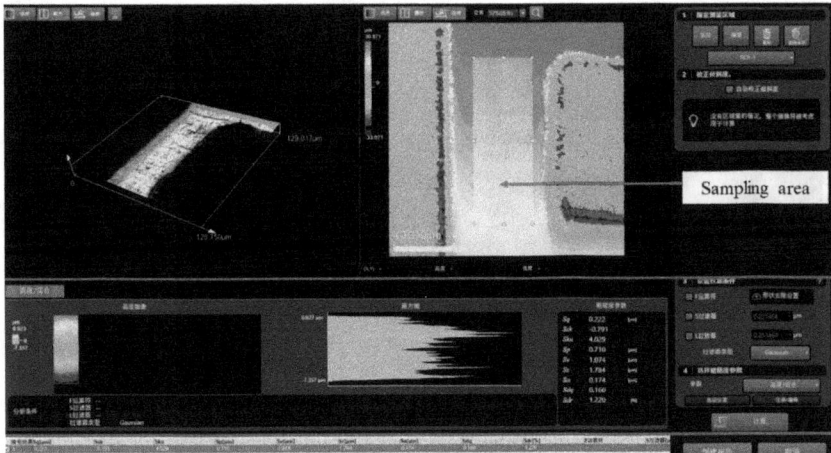

(a)

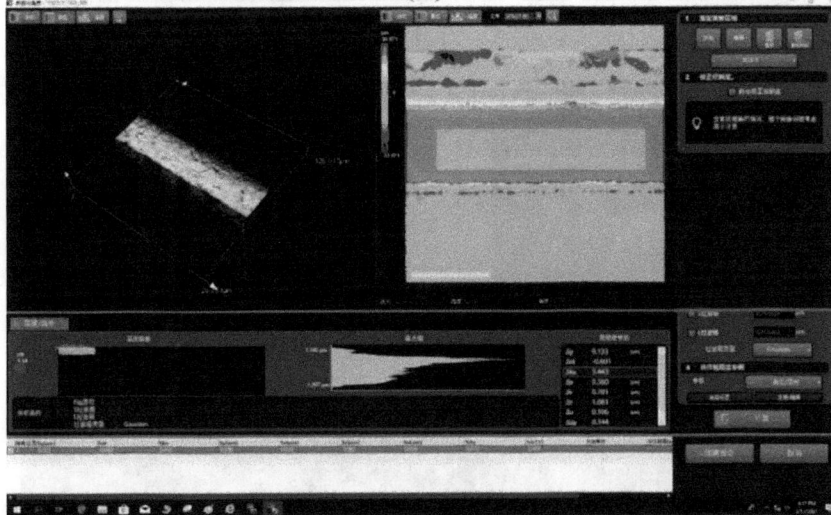

(b)

Figure 4.24. Surface roughness of pure molybdenum strip lines. (a) Turning section; (b) vertical segment.

laser cutting technology; (2) after copper is added via magnetron sputtering, the surface roughness of both the upper and side surfaces of the molybdenum is significantly decreased. This shows that the magnetron sputtering copper coating process used to improve the equivalent conductivity of materials not only has the great advantage of improving their inherent conductivity but also has a significant effect on reducing their surface roughness.

In view of the large differences in the roughness of each part of the strip line prepared by this process, the area proportion method was used to approximate the roughness of the whole surface of the magnetron-sputtered copper strip line on the

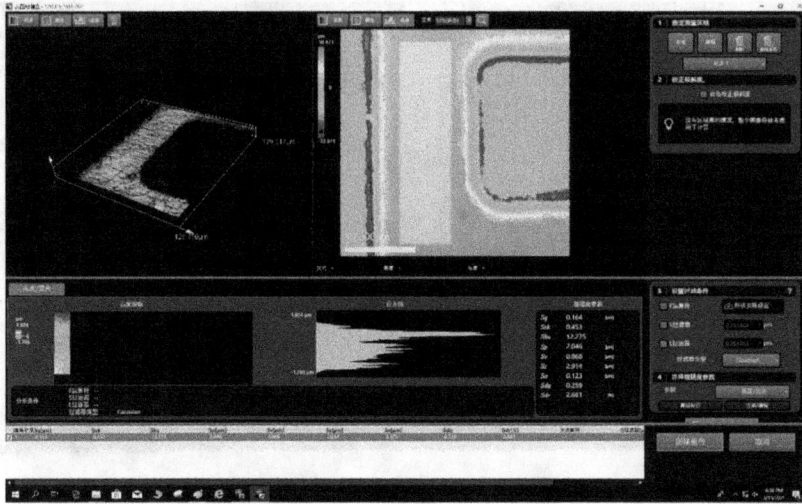

(a)

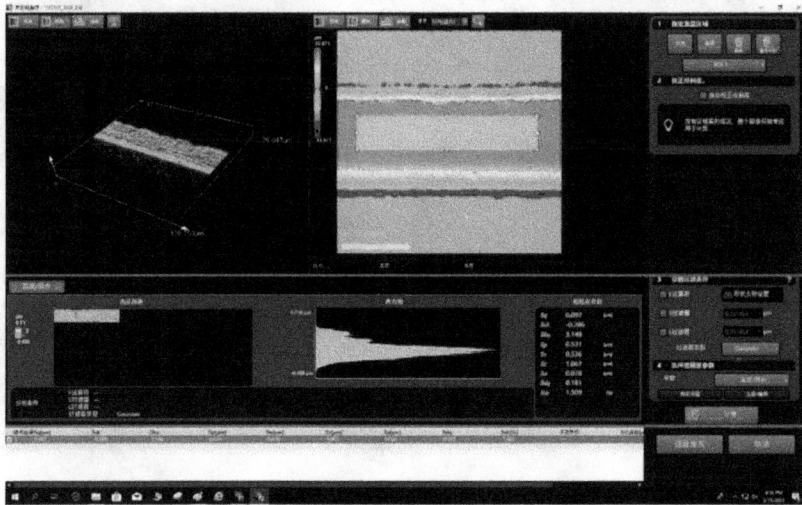

(b)

Figure 4.25. Surface roughness of a magnetron-sputtered copper layer on a strip line of molybdenum wire surfaces. (a) Turning section; (b) vertical segment.

surface of the molybdenum wire and the corresponding equivalent conductivity. Partial results are shown in tables 4.2 and 4.3.

According to this test, the full surface roughness value is 246 nm. Using formula (4.1), the equivalent conductivity at the frequency point of 38 GHz (the basis for the conductivity setting of the strip line in the simulation) was obtained as 2.94×10^7 S m^{-1}.

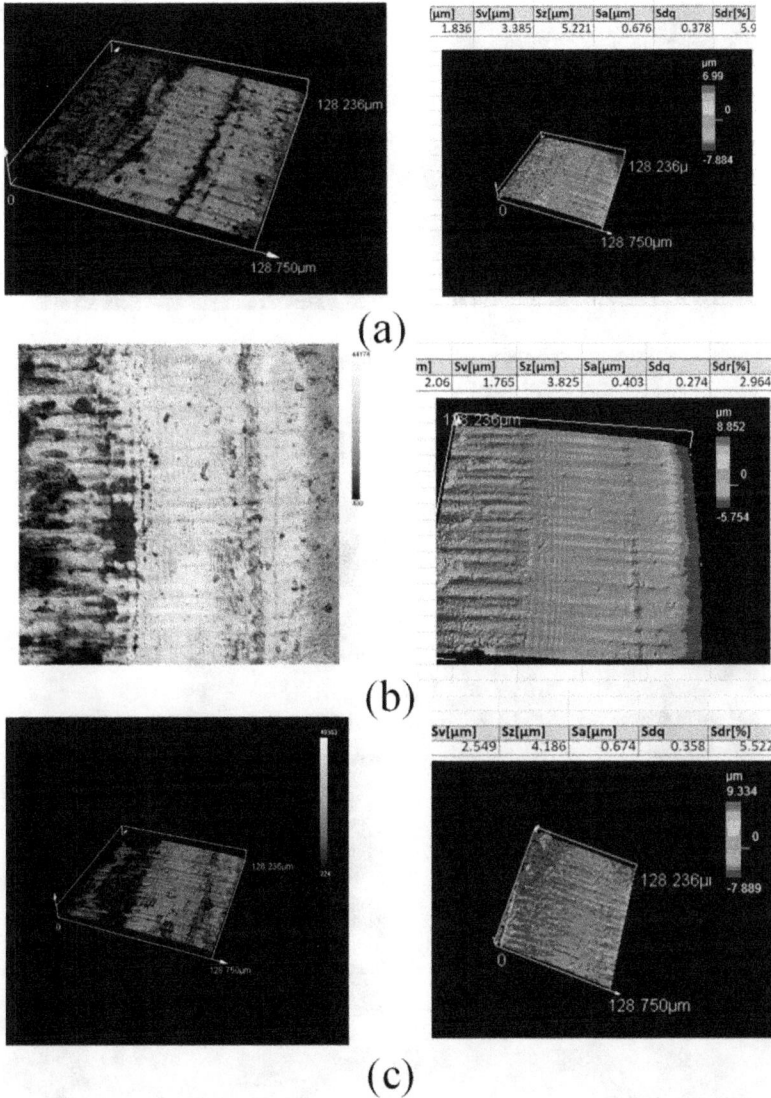

Figure 4.26. Side roughness of different areas of pure molybdenum strip lines. (a) Sampling area A; (b) sampling area B; (c) sampling area C.

4.6 Modified angular log-period folded waveguide SWS

For SWSs made from waveguides, micro/nano-CNC milling technology can provide better fabrication accuracy [11]. In this section, we introduce the fabrication of a modified angular log-period (MALP) meander waveguide SWS using micro-CNC milling technology. The material selected is bulk oxygen-free copper (OFC), brand TU1. Figure 4.28 shows a set of photographs of the SWS, and figure 4.29 shows measurements of some of the characteristic dimensions, including the width of the

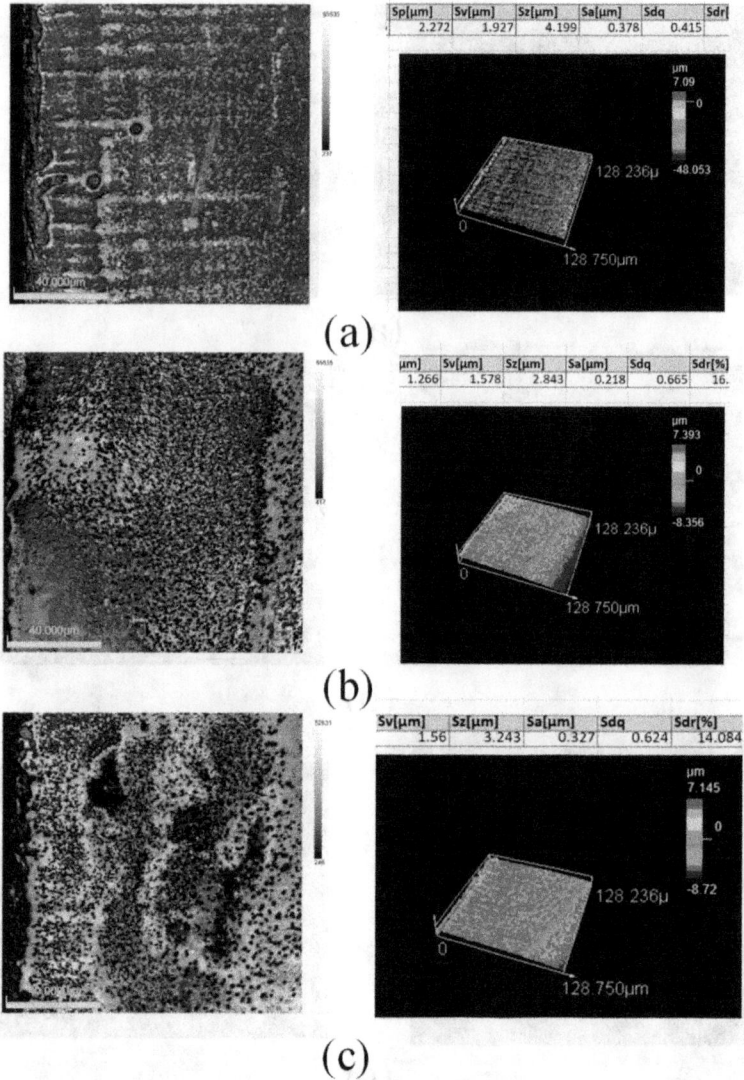

Figure 4.27. Side roughness of a magnetron-sputtered copper layer on a surface of molybdenum strip lines. (a) Sampling area A; (b) sampling area B; (c) sampling area C. Reproduced from [10]. CC BY 4.0.

Table 4.1. The roughness of strip lines made of two materials at different positions.

Material	Upper corner (nm)	Upper straight (nm)	Side A (nm)	Side B (nm)	Side C (nm)
Molybdenum	174	106	676	403	674
Molybdenum sputtered with copper	123	78	378	218	327

Table 4.2. The surface roughness of the magnetron-sputtered copper strip on the molybdenum wire surface and corresponding equivalent electrical conductivity.

Sample	R_S (nm)	σ_{ef} (10^7 S m^{-1}, $f = 38$ GHz)
Upper straight	78	5.34
Upper corner	123	4.66
Upper average (areal ratio 12:1)	76	5.31
Side average	307	2.41
Overall average (side–upper ratio 2.8:1)	246	2.94

Table 4.3. The average absolute dimensional tolerance of each section.

Section number	W_1	W_2	W_3	W_4	W_5
1	-0.87 μm	1.26 μm	-0.73 μm	-0.25 μm	-0.32 μm
2	-1.55 μm	1.38 μm	-1.47 μm	0.69 μm	-1.02 μm
3	-0.89 μm	0.63 μm	-1.05 μm	0.68 μm	-1.07 μm
4	-1.21 μm	1.42 μm	-1.92 μm	1.18 μm	-0.94 μm
5	-1.22 μm	1.44 μm	-1.50 μm	0.88 μm	-1.15 μm

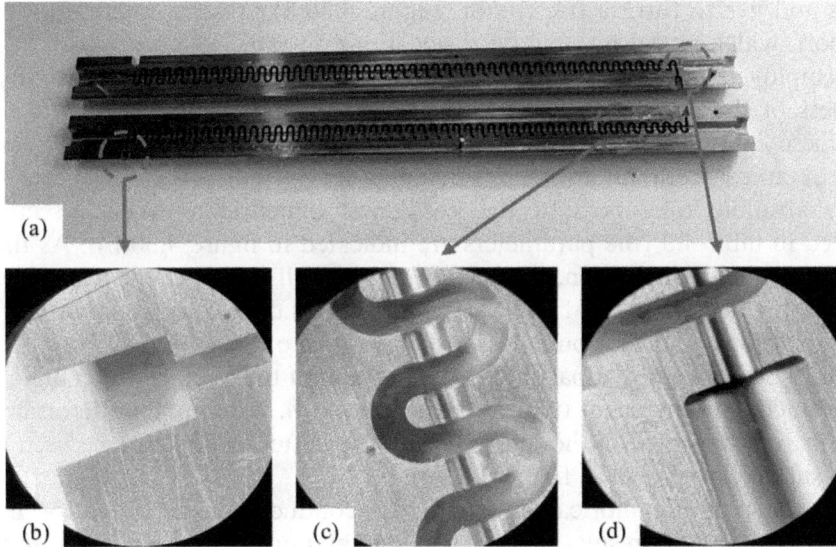

Figure 4.28. Photograph of the K_a-band MALP folded waveguide SWS. (a) General perspective; (b) output waveguide jack; (c) a meander path close to the input; (d) electronic injection channel entry. © [2023] IEEE. Reprinted, with permission, from [12].

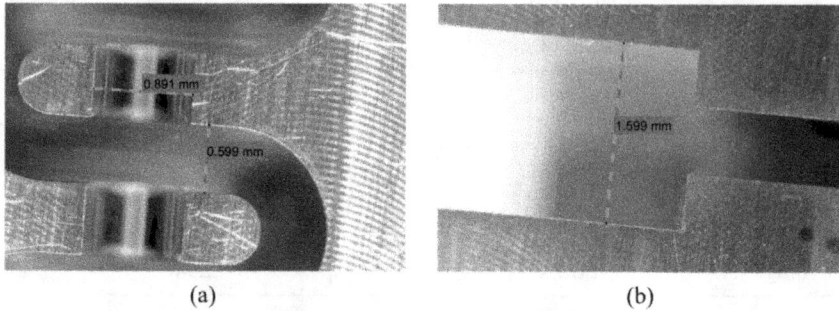

(a)   (b)

Figure 4.29. Partial feature size measurement of an SWS. (a) Waveguide width and electron beam channel diameter; (b) width of the output waveguide socket.

narrow side of the waveguide, the diameter of the electron beam channel, and the width of the output waveguide socket. It can be seen that the dimensional accuracy of the machined SWS is well controlled, and the measured dimensional errors are only a few microns.

4.7 Terahertz folded waveguide SWS

When the operational frequency increases to the terahertz band, the performance of an SWS becomes very sensitive to the fabrication tolerance. Even surface roughness at the micron level causes significant ohmic loss. In this section, we describe the fabrication of a folded waveguide SWS designed to operate at 0.66 THz. The SWS is produced by milling OFC via nano-computer numerical control (CNC). Typically, the high-frequency circuit is fabricated in two halves, which are aligned and secured by pins and jigs to form a full circuit. Figure 4.30(a) presents micrographs of the input port, which reveal no obvious dislocations or gaps.

We employed an optical profiler to measure the dimensions and bottom surface roughness of the circuit which were then used to update the simulation models. Using these figures, the dimensions of the single-period model were set to the mean values obtained from all measurement sites in the corresponding section. The average absolute tolerances in the horizontal dimensions of each section are presented in table 4.3 (the parameters are indicated in figure 4.30(b)). As the most frequency-sensitive dimension, the average depth (half-length of the wide side of the waveguide) is 141.345 μm (+1.345 μm); errors of less than 2 μm are observed for all other dimensions, and the roughness is in the range of 36.4–99.3 nm.

Figure 4.31(a) displays a partial 3D micrograph of the circuit, which has a length of 3 cm (including the beam tunnel extension length). No apparent deformation or oxidation was observed, although clear milling cutter marks can be seen on the bottom surface, as shown in figure 4.31(b).

A cold test of the fabricated circuit was conducted using a vector network analyzer (VNA); the results, along with simulations based on pre- and post-reconstructed models, are presented in figure 4.32(a). The initial conductivity of the OFC was set to 2×10^7 S m^{-1}, based on previous fabrication experience. As the measured S$_{21}$ was better than the simulated result, the conductivity was modified to

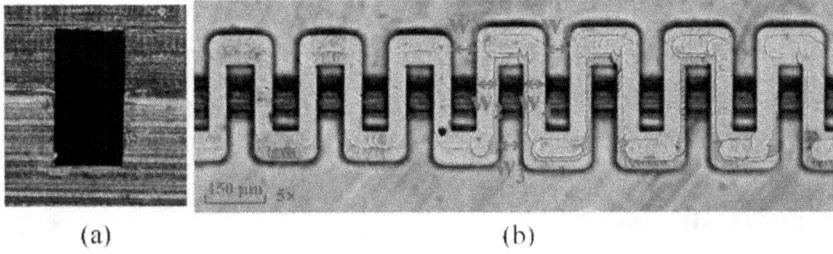

Figure 4.30. Micrographs of (a) the waveguide port and (b) the dimensionally varying part of the circuit. Copyright (2024) IEEE. Reprinted, with permission, from [14].

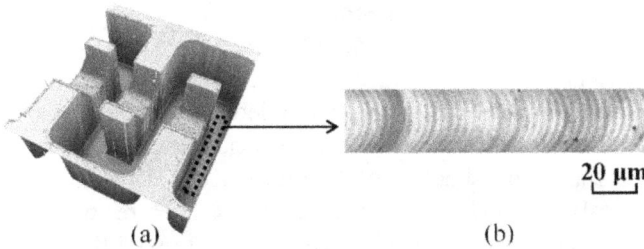

Figure 4.31. (a) Partial 3D micrograph of the circuit; (b) the bottom surface milling cutter marks. Copyright (2024) IEEE. Reprinted, with permission, from [14].

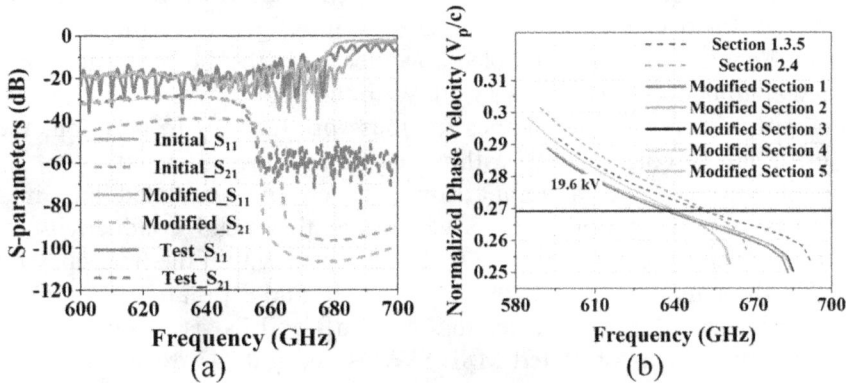

Figure 4.32. (a) Transmission characteristics of the cold test and the simulation. (b) Modified dispersion curves of each section. Copyright (2024) IEEE. Reprinted, with permission, from [14].

3.7×10^7 S m^{-1}, which gave consistent results between the simulation and the experiment. The transmission loss S_{21} in the frequency range of 630–650 GHz was -28 to -32 dB, the average transmission loss of the high-frequency circuit was only 0.96 dB mm^{-1} at the central frequency, and the reflection loss was less than -15 dB. The average surface roughness was approximately 45 nm, which represents a remarkably favorable outcome that surpasses recent reports of other terahertz circuits.

Figure 4.32(b) illustrates the modified dispersion (normalized phase velocity versus frequency) of the SWS in each section of the circuit. Machining errors, particularly positive errors of depth, led to a 12 GHz frequency shift with respect to the original synchronization voltage. Furthermore, the dispersion curve of the first section deviates from others with identical design dimensions, which will result in suboptimal modulation and bunching of the electron in this section at the center frequency.

4.8 Meander slot-line SWS

A meander slot line (MSL) proposed for a Q-band TWT was proved to have the benefits of a simple structure, wide bandwidth, and high beam–wave interaction impedance by simulation. The fabrication procedure used for this SWS is basically the same as that of the microstrip meander-line SWS [13].

Figure 4.33(a) shows the assembly model of a 29-period MSL-SWS with a pair of adapters, which was used for the transmission and particle-in-cell (PIC) simulations. The fabrication of the MSL-SWS was carried out as follows: first, the metal enclosure (copper, 5.8×10^7 S m^{-1}) was fabricated in two halves using CNC milling, and then the metal sheet (molybdenum, 1.8×10^7 S m^{-1}) was fixed to the bottom enclosure after being annealed to de-stress it. Here, we chose molybdenum rather than copper because copper is not rigid enough, so a thin copper sheet would have been severely deformed by laser cutting. A picosecond laser cutting machine (wavelength: 355 nm, average power: 10 W, pulse duration: 15 ps, pulse repetition: 600 kHz, spot diameter: 30–45 μm) was employed to engrave the meander slot line located at the center of the metal sheet, as shown in figure 4.33(b). At the end, the upper metal enclosure was assembled together with the bottom enclosure using screws and alignment pins (figure 4.33(c)). With the appropriate design, the laser can simultaneously cut multiple slot lines for mass production. In addition, the laser cutting error can be controlled to within 0.01 mm by measurement.

Figure 4.34(a) presents a comparison between the simulated and measured S-parameters of a 29-period MSL-SWS, where the metal conductivity used in the simulation was 1.48×10^7 S m^{-1}. As can be seen, within the frequency range of 35–44 GHz, the simulated S_{11} is better than -15 dB and S_{21} is approximately -2 dB, whereas the measured S_{11} is greater than -10 dB and S_{21} is about -3.4 dB. The insertion loss of the copper-plated MSL-SWS is about 0.043 dB mm^{-1}.

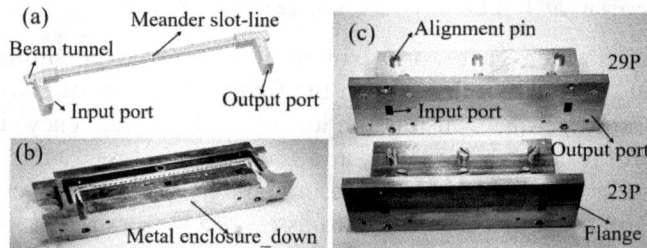

Figure 4.33. (a) Machined component processing diagram. (b) Photograph of the MSL-SWS. (c) U-shaped interdigital line array. © [2024] IEEE. Reprinted, with permission, from [13].

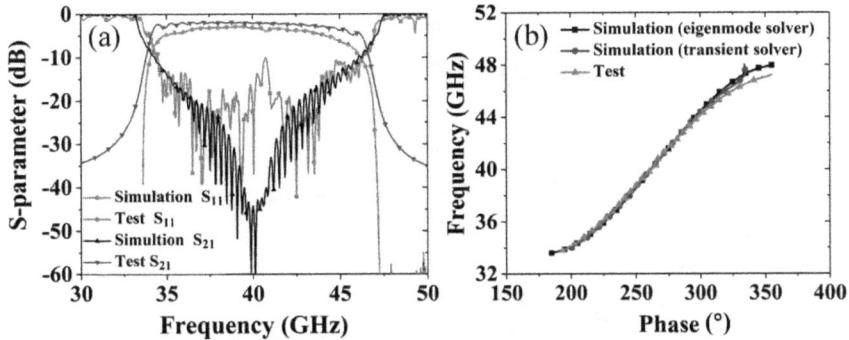

Figure 4.34. Comparison between simulated and tested (a) S-parameters for a 29-period MSL-SWS and (b) dispersion characteristics. © [2024] IEEE. Reprinted, with permission, from [13].

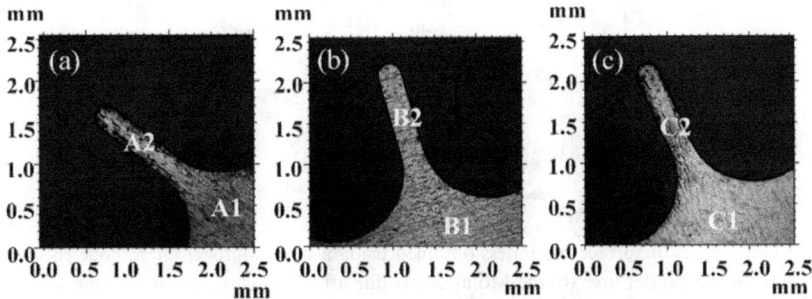

Figure 4.35. Optical microscopy images of U-shaped slot lines before pickling (a); after pickling (b); and with copper plating (c). © [2024] IEEE. Reprinted, with permission, from [13].

Figure 4.34(b) shows a comparison between the simulated and measured dispersion characteristic curves. Two MSL-SWSs with period numbers of 29 and 23 were used to calculate the simulated dispersion characteristic curves. As can be seen, the deviation between the simulated and measured dispersion characteristic curves is about 0.447%.

In order to find the reason for the differences between the simulated and measured S-parameters, the surface roughness of the fabricated MSL-SWS was investigated. Figure 4.35 shows one tooth of the MSL-SWS under a microscope. The contaminants and black oxides adhering to the surface of the directly cut slot line in figure 4.35(a) were successfully eliminated after washing with a 5% hydrochloric acid solution (figure 4.35(b)). After copper plating with a thickness of 3–5 μm was added, as depicted in figure 4.35(c), the surface of the structure appeared more homogeneous. Consequently, laser confocal microscopy was employed to measure the size of the surface roughness at two sites in two cases, as shown in figure 4.36. The results of the study found that the surface of the structure was more inhomogeneous at the laser spot path (A2, B2), which could be effectively improved by acid cleaning and copper plating.

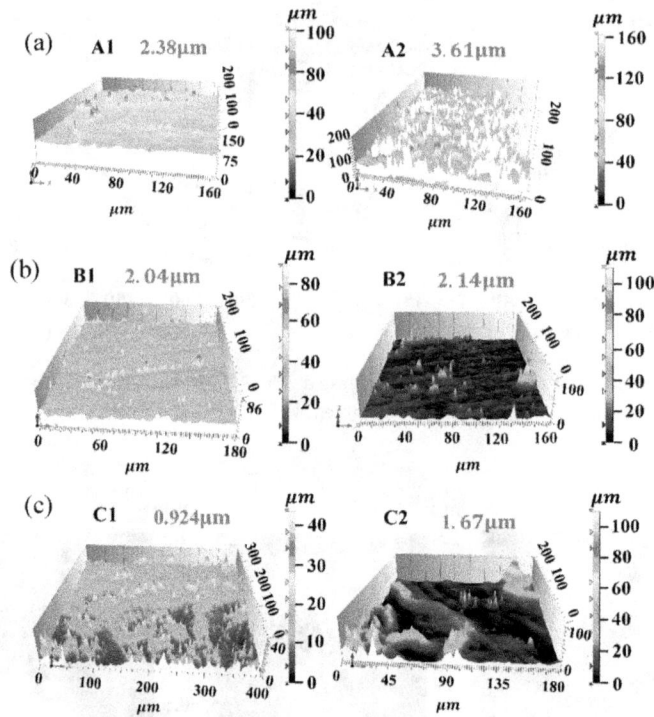

Figure 4.36. Test images of surface roughness obtained using a laser confocal microscope (LSM 800) at the surface of (a) the Mo meander line; (b) the Mo meander line after acid cleaning; and (c) the Mo meander line sputtered with copper film. © [2024] IEEE. Reprinted, with permission, from [13].

In addition, the molybdenum sheet was laser cut into two separate pieces during manufacture, which were combined to form an interdigital line. As a result, the two pieces were relatively displaced, which caused a tiny reflection of S_{11} at a single frequency point.

Given the previous discussion, changing the material's conductivity by plating it with copper is the best way to reduce insertion loss. In addition, assembly faults can be avoided, which decreases reflection losses.

References

[1] Sumathy M, Augustin D, Datta S K, Christie L and Kumar L 2013 Design and RF characterization of W-band meander-line and folded-waveguide slow-wave structures for TWTs *IEEE Trans. Electron Devices* **60** 1769–75

[2] Ryskin N M, Torgashov R, Starodubov A, Rozhnev A, Serdobintsev A, Pavlov A, Galushka V, Bessonnov D, Ulisse G and Krozer V 2021 Development of microfabricated planar slow-wave structures on dielectric substrates for miniaturized millimeter-band traveling-wave tubes *J. Vac. Sci. Technol.* B**39** 013204

[3] Starodubov A V, Serdobintsev A A, Pavlov A M, Galushka V V, Ryabukho P V and Ryskin N M 2018 A novel approach to microfabrication of planar microstrip meander-line slow

wave structures for millimeter-band TWT *Presented at 2018 Progress in Electromagnetics Research Symp. (Piers-Toyama)*

[4] Bai N F, Feng C, Liu Y T, Fan H H, Shen C S and Sun X H 2017 Integrated microstrip meander-line traveling wave tube based on metamaterial absorber *IEEE Trans. Electron Devices* **64** 2949–54

[5] Sengele S, Jiang H, Booske J, Kory C, Daniel W and Ives R 2009 Microfabrication and characterization of a selectively metallized W-band meander-line TWT circuit *IEEE Trans. Electron Devices* **56** 730–7

[6] Wang S, Aditya S, Xia X, Ali Z and Miao J 2018 On-wafer microstrip meander-line slow-wave structure at Ka-band *IEEE Trans. Electron Devices* **65** 2142–8

[7] He T *et al* 2018 Study on silicon-based conformal microstrip angular log-periodic meander line traveling wave tube *Prog. Electromagn. Res.* M **75** 29–37

[8] He T, Li X, Wang Z, Wang S, Lu Z, Gong H, Duan Z, Feng J and Gong Y 2020 Design and cold test of dual beam azimuthal supported angular log-periodic strip-line slow wave structure *J. Infrared Millim. Terahertz Waves* **41** 785–95

[9] Li X, He T, Wang H, Chen Z, Wang Z, Gao Z, Duan Z, Wei Y and Gong Y 2018 Microstrip angular log-periodic slow wave structure on quartz substrate with coaxial input/output coupler *J. Eng.* **2018** 692–7

[10] Wang H, Wang S, Wang Z, Li X, He T, Xu D, Duan Z, Lu Z, Gong H and Gong Y 2021 Study of an attenuator supporting meander-line slow wave structure for Ka-band TWT *Electronics* **10** 2372

[11] Xu D *et al* 2020 Theory and experiment of high-gain modified angular log-periodic folded waveguide slow wave structure *IEEE Electron Device Lett.* **41** 1237–40

[12] Xu D, Wang S, Lu C, He T, Wang Z, Lu Z, Gong H, Duan Z and Gong Y 2023 Demonstration of a modified angular log-periodic folded waveguide traveling wave tube at Ka-band *IEEE Trans. Electron Devices* **70** 1323–9

[13] Wang Y, Wang S, Dong Y *et al* 2024 Investigation of a novel planar meander slot-line slow wave structure *IEEE Electron Device Lett.* **45** 476–9

[14] Guo J *et al* 2024 Novel low-loss 0.65-THz multisectional folded waveguide high-frequency circuit *IEEE Trans. Electron Devices* doi: 10.1109/TED.2024.3452703

Chapter 5

Fixtures and assembly

Generally, a traveling-wave tube (TWT) consists of tens of components that realize the energy exchange between the electron beam and the electromagnetic wave. These components have to be assembled accurately to operate correctly; therefore, fixtures have to be designed and manufactured. In this chapter, fixtures for two different planar TWTs are introduced, i.e. the angular log-period meander strip-line TWT and the modified angular log-period (MALP) folded waveguide TWT [1–5].

5.1 Angular log-period meander strip-line TWT

5.1.1 Angular radial electron optical system

5.1.1.1 Electron gun
The key components of the angular radial electron-beam gun include the gun shell, the hot wire assembly, the cathode head assembly, the control electrode, the anode, and the gun end cap, among which the gun end cap is used to provide an interface with the slow-wave structure (SWS) assembly (or the electronic beam tunnel assembly). See figure 5.1 for an assembly diagram of the flow tube of the angular radial electron-beam gun.

The assembly of the electron gun needs to go through a series of complex technological processes, such as the metallization of the sealing positions of the ceramic parts of the gun shell, the ceramic–metal sealing between the metallized ceramic of the gun shell and the electrode lead ring and the gun end cap, the processing, heat treatment, salt dipping (in an electron-emitting material) of the cathode head, and the welding process between the hot plate cylinder and the support cylinder.

The following sections elaborate the design and experimental study of the flow tube, including the electron gun, the flow channel, and the overall tube assembly.

doi:10.1088/978-0-7503-5452-3ch5

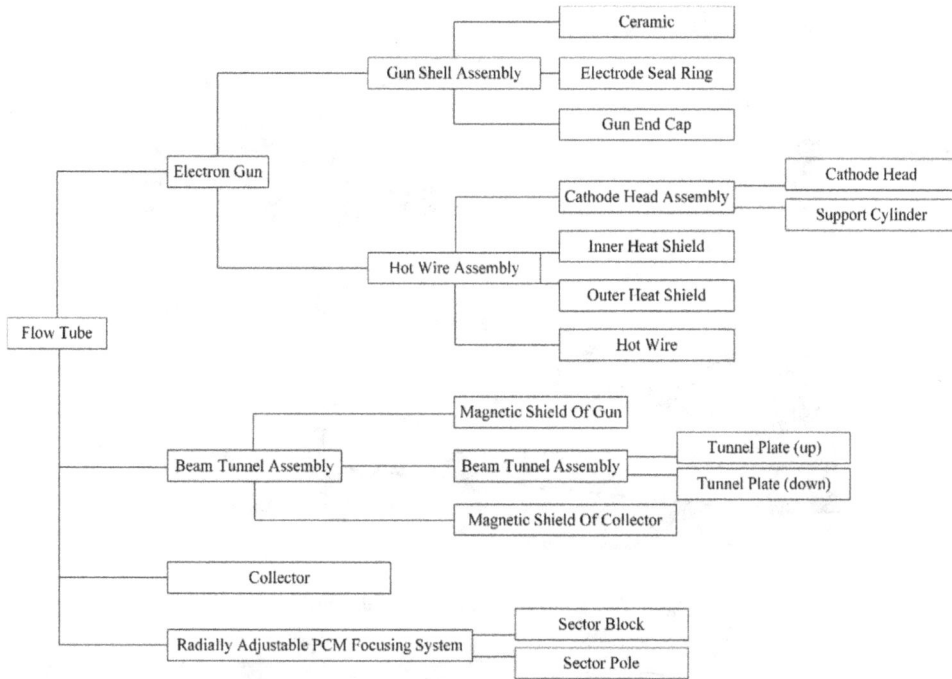

Figure 5.1. Electron beam flow tube assembly tree diagram.

5.1.1.1.1 *Structural design*

Figure 5.2 shows a 3D design drawing of the cathode head parts, which strictly guarantee the morphology of the emission surface, namely the radii of the arc surfaces with thicknesses of 0.6 mm and 14.5 mm. The part below the arc emission surface is rectangular, which is convenient for processing and assembly. According to the above design principles, the electron gun in this paper uses mature designs for the tubular shell, hot wire assembly, and supporting parts except for the supporting cathode head and the beam-forming electrode of the current process line.

The cathode head assembly is shown in figure 5.3. The cathode head support cylinder has two functions: (1) to support the cathode head and (2) to heat the cathode head using a heating wire in the cavity under the cathode head support cylinder (which also contains an insulating filler).

Figure 5.4 shows the design diagram of the beam-forming electrode. In principle, the design of the engineering drawing should be consistent with the simulation model. However, because the height of the angular dimension of the angular bevel is less than 0.2 mm, the current electric discharge machining (EDM) technology cannot form such a shape. Simulation calculations show that the angular bevel can be degraded into a horizontal step, so only the axial bevel at an angle of 80.8° is retained. The other structural dimensions adjacent to the cathode head are retained. The disc at the bottom of the beam-forming electrode can be adjusted according to the specific assembly requirements. The structure of this part, which plays the role of supporting flange, has no influence on the formation of the electronic beam.

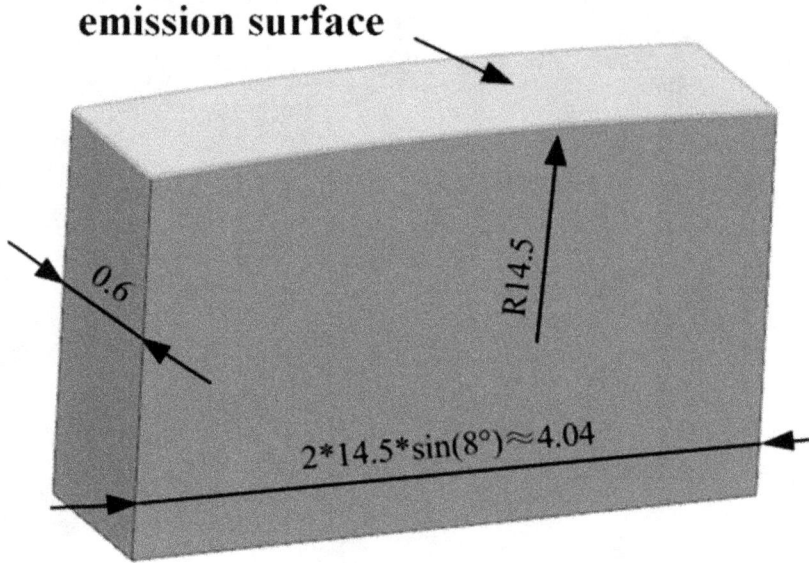

Figure 5.2. Cathode head and key dimensions.

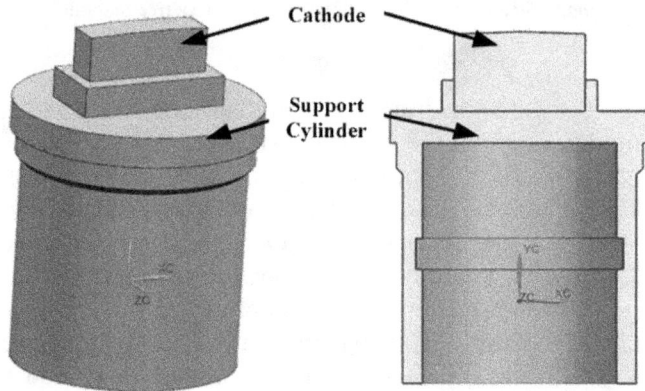

Figure 5.3. Cathode head components.

Figure 5.5 shows the third key part of the electron gun used to form the electron beam, namely the anode. The anode of a traditional helix TWT is a separate part, as shown in figure 5.6. It is welded to the input cap of the SWS by laser spot welding or a vacuum brazing process and acts as both an anode and an electron-beam drift tunnel. After the input end cap is welded to the end cap of the electron gun shell (using argon arc welding, electron-beam welding, etc.), the complete structure of the electron gun is formed, consisting of the three key parts of the cathode, the beam-forming electrode, and the anode. Meanwhile, the components of the SWS are also assembled.

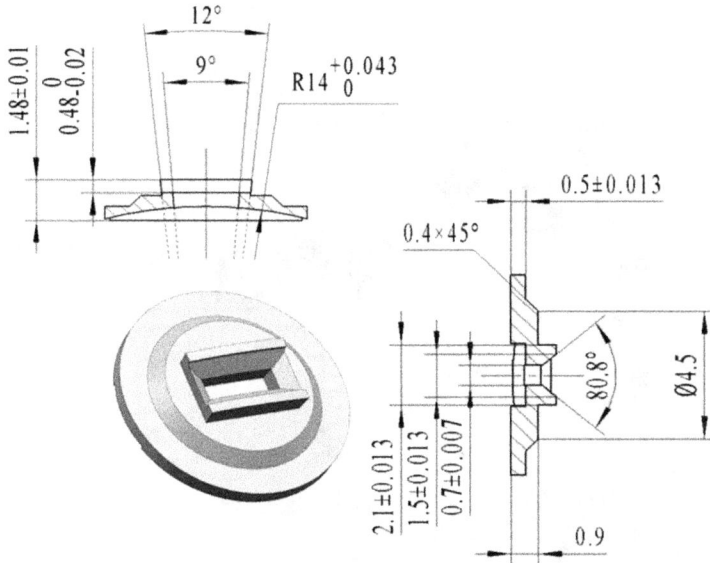

Figure 5.4. Engineering design of the beam-forming electrode.

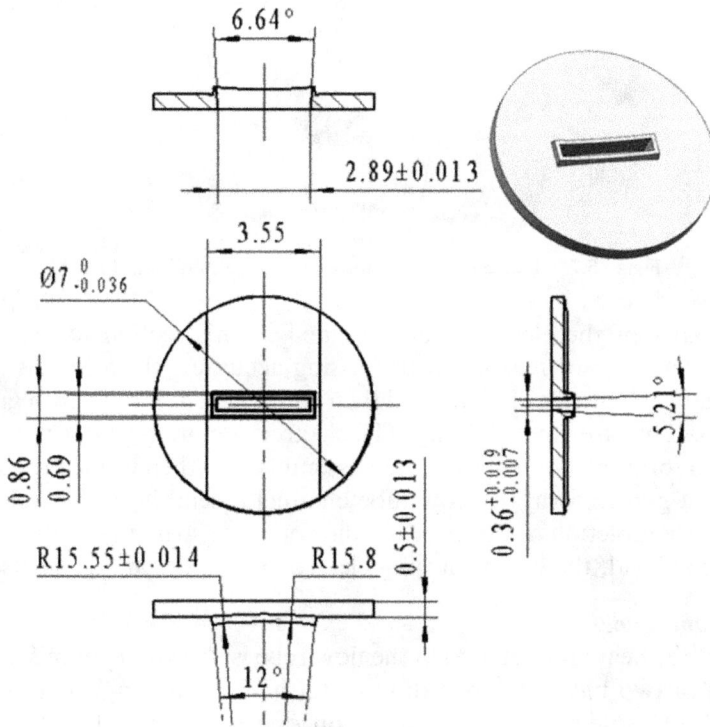

Figure 5.5. Engineering design diagram of the anode.

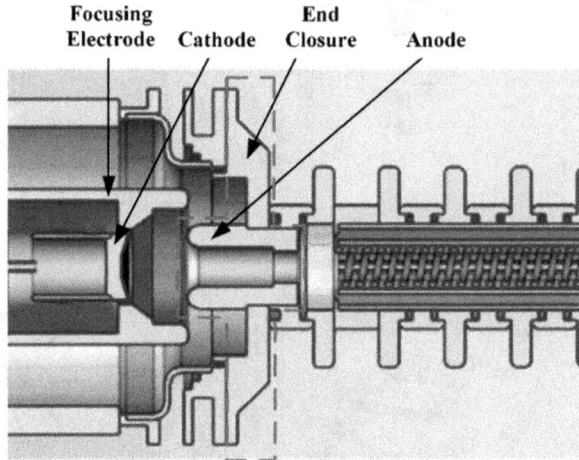

Figure 5.6. Anode assembly of a traditional helix TWT.

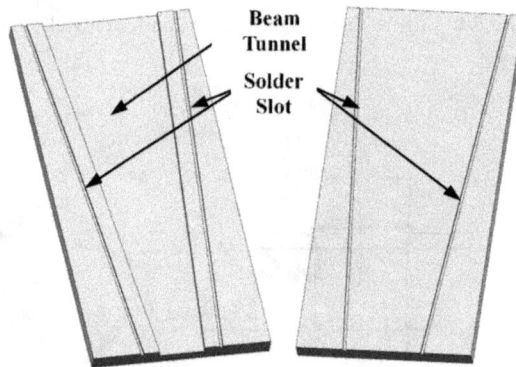

Figure 5.7. Three-dimensional model of the electron-beam tunnel.

The key parts of the electron gun designed in this section have a non-axially symmetric structure. For reasons of processing accuracy, the anode is made into a part with a flange diameter of 7 mm and then embedded in the electron gun magnetic screen and fixed by laser spot welding. The electron gun magnetic screen is welded to the channel to form the flow tube channel assembly and then welded to the end cover of the electron gun to form the flow tube channel assembly of the welded electron gun. After the completion of this process, the complete structure of the electron gun, i.e. the cathode head, the beam-forming electrode, and the anode, is also complete.

5.1.1.1.2 Beam tunnel
The design of the beam tunnel used in the flow tube is shown in figure 5.7. The tunnel is composed of two nickel–copper alloy metal plates. The angular radial electron-beam tunnel and solder slots are formed on one of the metal plates, and silver–copper-28 solder wire with a diameter of 0.3 mm is placed on the slots. The role of

the solder slots is to be filled with solder, thereby ensuring that the welds on both sides of the electronic beam are airtight after welding.

We now weld the electron gun magnetic screen and the collector magnetic screen at the corresponding positions on both sides to complete the welding of the channel assembly. The flange sizes of the electron gun magnetic screen and the collector magnetic screen are consistent with the corresponding electron gun end cover and the collector end cover. The rectangular hole that forms a butt joint with the channel component is machined in the central position, and the channel component is formed by welding using a suitable solder.

5.1.1.1.3 *Flow tube assembly design*

The schematic diagram is shown in figure 5.8, including the collector for the angular radial electron beam.

During the assembly of the flow tube, some tube components were photographed, such as the cathode head assembly of the welded beam-forming electrode, the gun shell assembly, and the completed electron gun assembly, as shown in figures 5.9(a)–(c), respectively. The bar gap corresponding to the tunnel can be clearly seen at the entrance to the inner conductor of the collector assembly in the tunnel and collector assembly shown in figure 5.9(d).

The final assembly process of the whole flow tube is shown in figure 5.10, where figures 5.10(a)–(c) show the components of the flow tunnel component after the argon arc welding of the collector magnetic screen end and the collecting component, the components after the argon arc welding of the electron gun magnetic screen end and the electron gun component, and after the argon arc welding of the electron gun and exhaust tube assembly, respectively.

After the completion of each welding procedure, vacuum leakage detection should be carried out, and the subsequent procedure should only be undertaken if the inspection is successful. The vacuum leakage detection instrument used is a helium mass spectrometer, and the qualified leakage rate is 3e-10 Pa m^3 s^{-1} required by the cover inspection (i.e. the whole tube is immersed in a helium atmosphere). For more information on vacuum leakage detection, see reference [3].

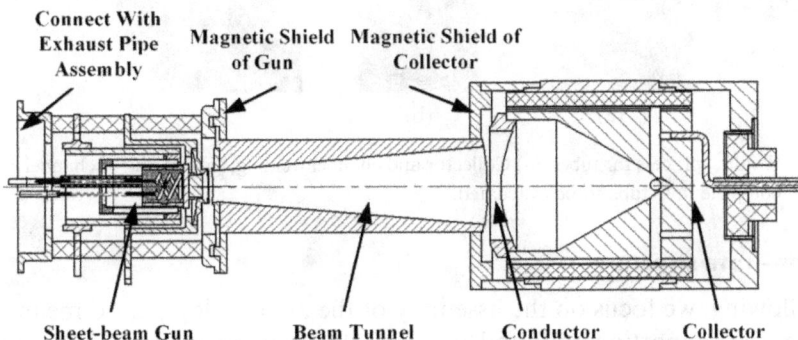

Figure 5.8. Schematic diagram of the assembled flow tube.

Figure 5.9. Electron gun components, channel components, and collector components. (a) The cathode head assembly with the welded beam-forming electrode; (b) gun shell assembly; (c) an assembled electron gun (sealed in glass); and (d) beam tunnel and collector assembly. © [2017] IEEE. Reprinted, with permission, from [4].

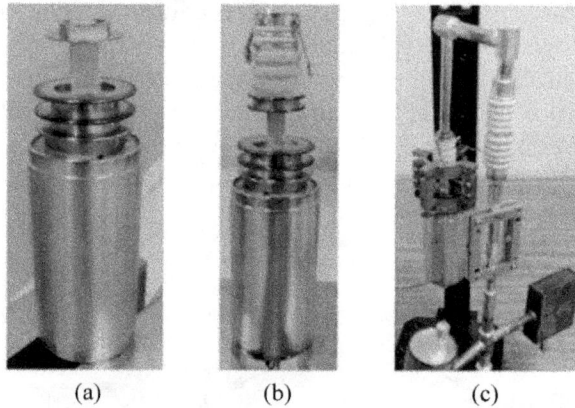

Figure 5.10. Welding process for tubes. (a) Collector and channel welding; (b) welding of channel and electron gun; and (c) complete flow tube to be evacuated.

5.1.2 Slow-wave structure

In the following, we focus on the assembly of the angular log-period meander slow-wave line, the supporting rod, and the assembly process used in SWS research in detail.

The refractory metal material used is a molybdenum–rhenium alloy with a sheet thickness of 0.1 mm. Figure 5.11 shows an angular log-period meander strip line with a line width of 0.075 mm processed by a picosecond laser-cutting machine. The initial radius is 16 mm and the angle is 4°.

There are fewer gradient sections at the input end in the structure actually processed. The main reason for this is that the concentric assembly structure between the key components of the angular logarithmic radial beam TWT reduces the radius of the emission surface of the cathode head accordingly. At the same angle, the radius decreases along with the emission surface. A failure to obtain the same current leads to an increase in the emission density of the current at the emission surface, thus increasing the cathode preparation burden.

Figure 5.12 shows the input–output coupling structure. The main adjustment is that instead of using a two-period strip-line jump matching structure at the input end, the strip-line jump matching structure is only used at the output end.

The angular log-period meander slow-wave line shown in figure 5.11 should not be hand-held in the assembly process because the angular logarithmic meander line has a 45° chamfered angle at the right-angle bend, which weakens the connection strength at the corner. Hand-held operation is especially likely to cause irreversible deformation of the meander line.

In view of the above, it is necessary to fix the strip-line slow-wave line onto the fixture when assembling and thus protect the shape of the slow-wave line from deformation due to accidental force during the processes of assembly and welding.

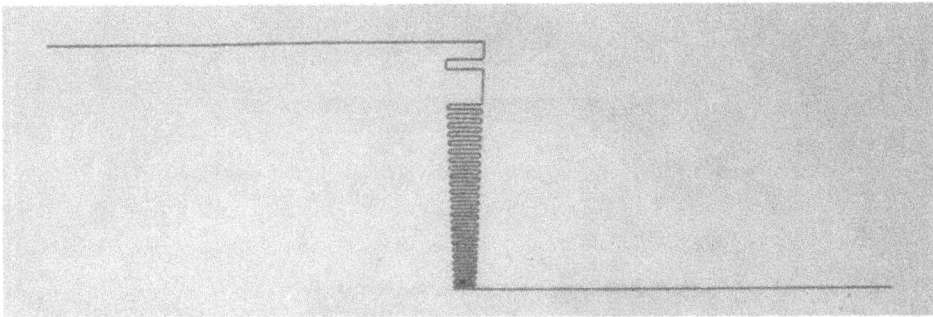

Figure 5.11. Angular log-period meander slow-wave line produced by laser cutting.

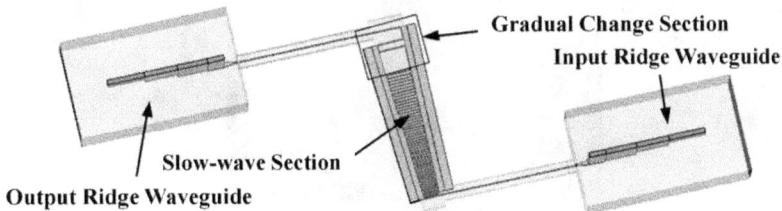

Figure 5.12. Angular log-period meander slow-wave line with input/output adapters.

Figure 5.13 shows the design of the fixture used to fix the meander strip line. The function of the fixture is to assemble the meander strip line on the support block of the slow-wave line and ensure a consistent assembly position. The assembly steps used for this fixture are as follows:

(1) Place a piece of meander strip-line support block into the slot on the support base, ensuring the depth of the slot is consistent with the thickness of the slow-wave line support piece.

(2) Place the meander strip line in the meander strip-line limiting slot of the supporting base (see figure 5.14). Place the second meander strip-line supporting block, and fix the two meander strip-line support blocks using fastening strips.

(3) Use screws and nuts to fix the two meander strip-line supporting blocks pressed by fastening strips.

(4) Remove the fastening strip, and take out the support base together with the fastened slow-wave line support piece and the meander strip line.

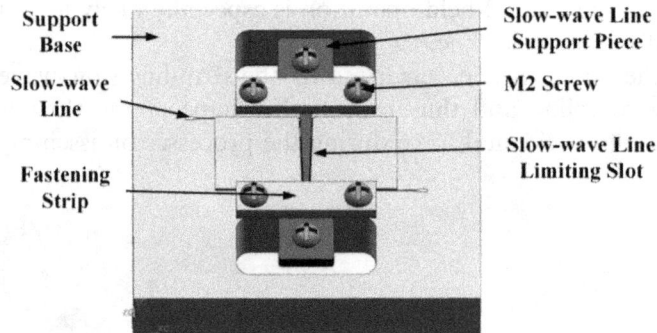

Figure 5.13. Assembling a fixture and a meander slow-wave line.

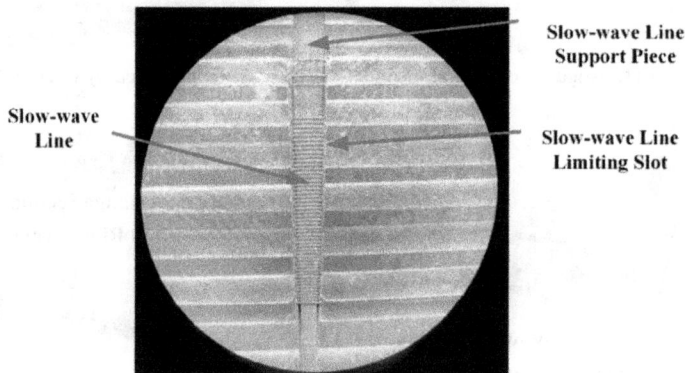

Figure 5.14. Photomicrograph showing the installation of the meander strip line in the meander strip-line limiting slot.

The welding between the strip slow-wave line and the T-shaped gripper rod must not cause a short circuit between adjacent solder joints, so it is necessary to strictly control the amount of titanium–silver–copper solder used for coating. Our current process uses a clipped pencil as a coating tool, which generally results in overcoating. Here, a 0.2 mm solder wire was used as a solder coating tool to apply solder to the straight sections of the strip slow-wave line one by one under the microscope.

After solder is applied to the straight sections of the meander strip line one by one, it is placed in the welding fixture to form a butt joint with the T-shaped gripper rod. A design drawing of the assembly fixture is shown in figure 5.15. The assembly steps are as follows:

(1) Place the zoned meander strip line (including the supporting fixture) coated with solder in the limiting slot of the welding base, and secure it with fastening strips using screws.

(2) Place the T-shaped gripper on both sides of the meander strip line coated with solder, ensuring that the T-shaped gripper rod and slow-wave line are not in contact.

(3) Place the clamping rod limit fixed slider in the welding base guide groove, and, using the screw, preliminarily define its position; however, do not fasten it at this time, so that the clamping rod limit fixed slider can slide along the guide groove.

(4) Place the T-shaped clamping rod in the gap of the clamping rod limit fixing slider, as shown in figure 5.16, and push the T-shaped clamping rod on both sides of the slow-wave line to the slow-wave line through the sliding block at the same time. After full contact is achieved, tighten the screws to fix the assembly.

Figure 5.17 shows a photo of the assembly parts to be welded. As shown in the figure, the solder between the T-shaped gripper rod and the meander strip line is evenly distributed in and around the contact surfaces of the two. The welding temperature of the titanium–silver–copper solder used is 800 °C, and the welding is done in a vacuum furnace.

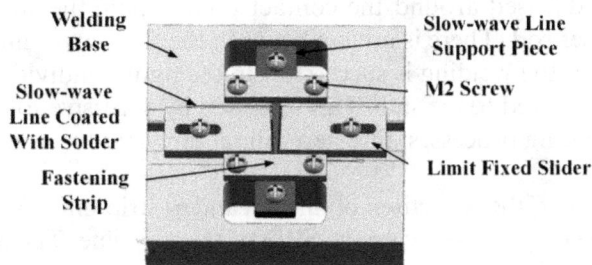

Figure 5.15. Welding fixture design of SWS.

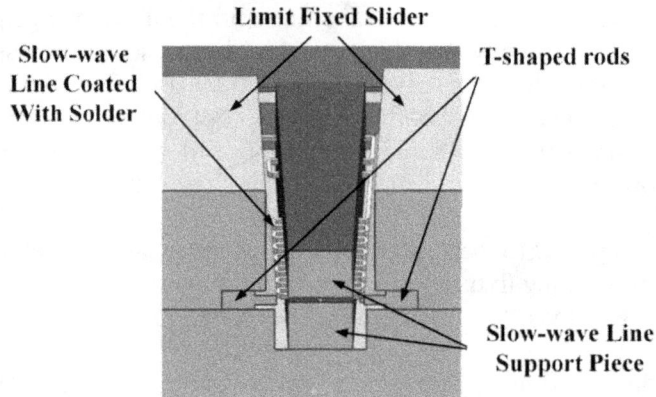

Figure 5.16. Cross-section of the slow-wave welding fixture.

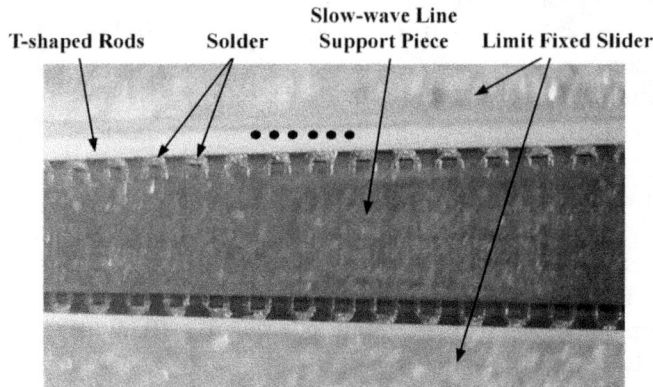

Figure 5.17. SWS to be welded (partial photo).

The temperature setting curve of the vacuum furnace is shown in figure 5.18. Starting from room temperature (20 °C), the temperature rises to 260 °C at the rate of 8 °C min^{-1} and then rises to 650 °C, 720 °C, and 800 °C at rates of 10 °C min^{-1}, 15 °C min^{-1}, and 16 °C min^{-1}, respectively. After being held at 800 °C for 5 min, the vacuum furnace enters a natural cooling stage until it reaches room temperature.

The welded SWS component is shown in figure 5.19. In the figure, the solder has been melted and diffused around the contact point where the strip slow-wave line contacts the gripper rod. There is no gap between the slow-wave line and the gripper rod, indicating that the welding is successful. In the figure, individual contact points of the solder have flowed to the step. This is caused by excessive solder point coating. In future point-coating processes, this step will require care to avoid the use of excess solder.

Figure 5.20 shows the structure of the meander strip-line SWS assembly, and figure 5.21 is a line diagram showing the SWS of the strip line. The meander strip-line SWS is assembled in the clamping groove of the clamping rod arranged on the bottom plate, and the cover plate is added to complete the overall assembly of the SWS.

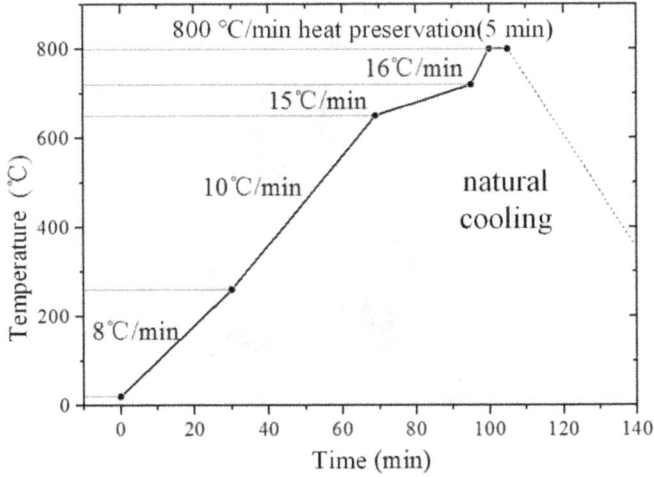

Figure 5.18. Temperature setting curve of the titanium–silver–copper active solder vacuum furnace.

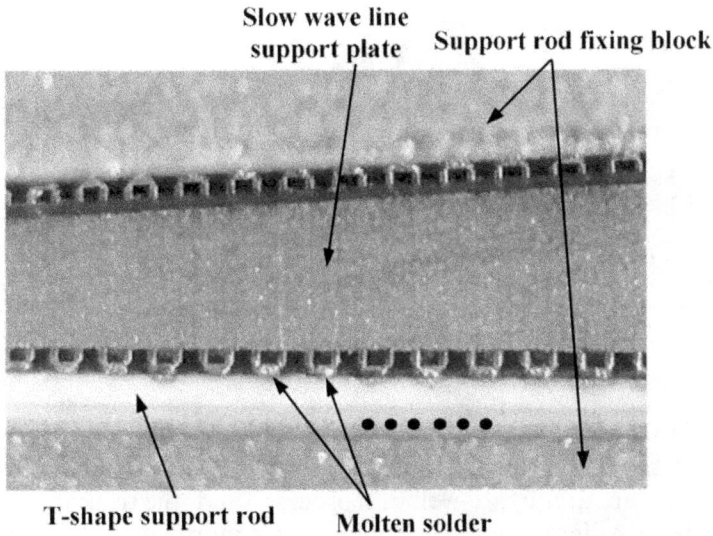

Figure 5.19. SWS after welding (partial).

Molybdenum–rhenium alloy, oxygen-free copper, and nickel–copper alloy (Monel) cannot be directly welded through spot welding (resistance welding). In addition, due to the narrow width of the ladder ridge (0.3 mm–0.4 mm) and the inconvenience of solder placement and solder flow problems, it is not possible to use solder welding. To address this issue, a 0.07 mm thick platinum sheet with the same area as that of the first section of the step ridge in contact with the meander strip line is placed between the step ridge and the strip slow-wave line to provide a welding transition between the two and thus permit a spot-welding method to be used.

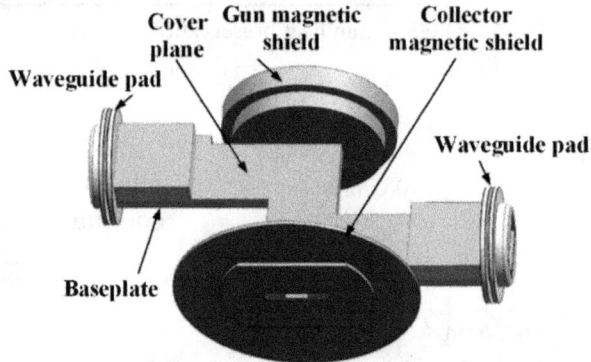

Figure 5.20. SWS design model.

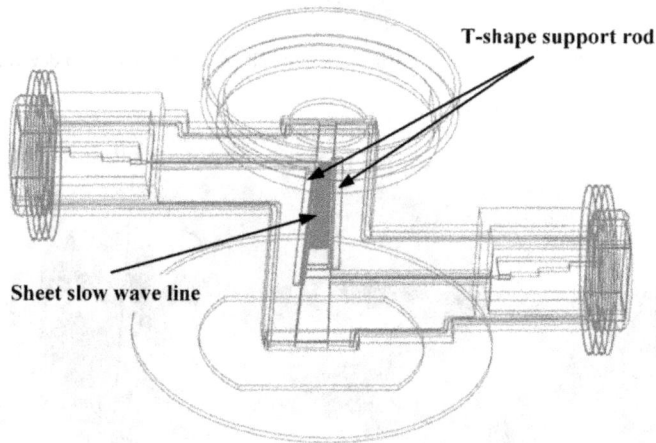

Figure 5.21. SWS design model (line drawing).

As shown in figure 5.22, there is one more collector connecting ring part in this picture than in the design model. This part and the collector magnetic screen are assembled using an argon arc-welding process, and the leak detection rate is measured using a helium mass spectrometer to confirm that the seal quality has reached the expected design goal. It can also be seen that the solder diffusion areas of the two waveguide pads, the electron gun magnetic screen, and the connected channel are large, so the solder consumption can be reduced in the subsequent assembly of the whole tube. After a comparison test, the solder consumed in welding the waveguide pads and the magnetic screen can be determined.

A photo of the SWS assembly (excluding the electron gun magnetic screen and the collector magnetic screen) is shown in figure 5.22.

5.1.3 Radial electron-beam magnetic focusing system

Before the flow rate test can take place, the radial focusing system needs to be assembled. Here, a radial tunable periodic cusped magnetic (PCM) focusing system

Cover plane **Gun magnetic shield** **Collector magnetic shield**

Waveguide pad

Waveguide pad

Baseplate

Collector connection ring

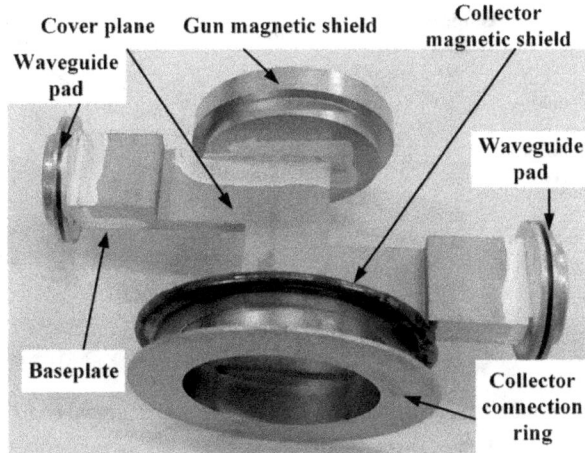

Figure 5.22. Photo of the SWS welding.

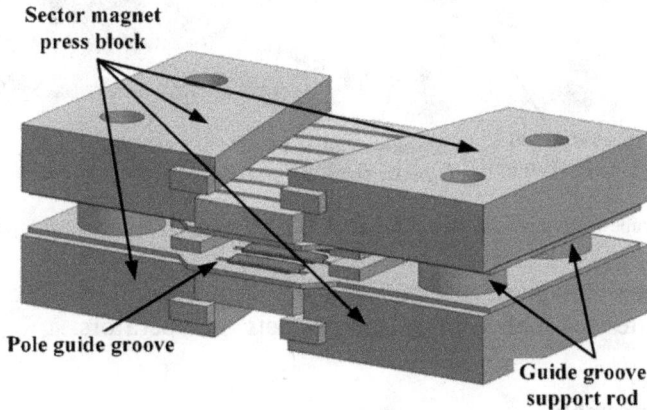

Sector magnet press block

Pole guide groove

Guide groove support rod

Figure 5.23. Design of the fixture for the radially tunable PCM focusing system.

is designed (see figure 5.23 for the design schematic diagram). Figure 5.23 shows the assembly fixture design of the magnetic system, and figure 5.24 shows the physical view. The roles of the pole piece guide slot are:

(1) to limit the adjustable rotational range of the pole piece to an arc interval equal to its radius, and

(2) to keep the fan-shaped magnetic block on a plane.

The assembly of the radially tunable PCM focusing system can be completed by engaging the sector with the beveled edge of the sector block and fastening it with the (axial) opposite sector block together with the guide groove support bar using screws. All parts of the magnetic system assembly jig are made of duralumin.

Figure 5.25 shows the flow tube equipped with the magnetic focusing system. The shell of the electron gun has been treated with adhesive insulation to prevent short

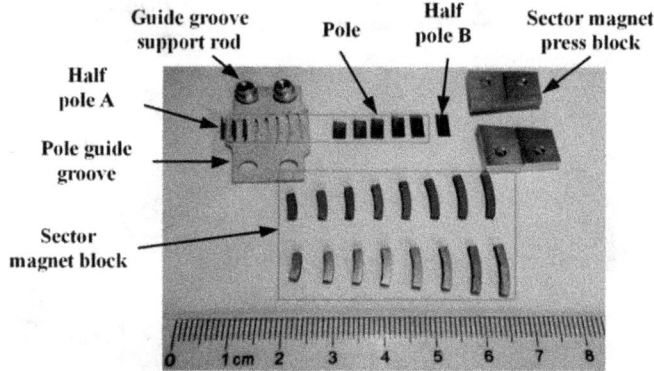

Figure 5.24. Parts of the radial tunable PCM focusing system and assembly fixtures. © [2017] IEEE. Reprinted, with permission, from [4].

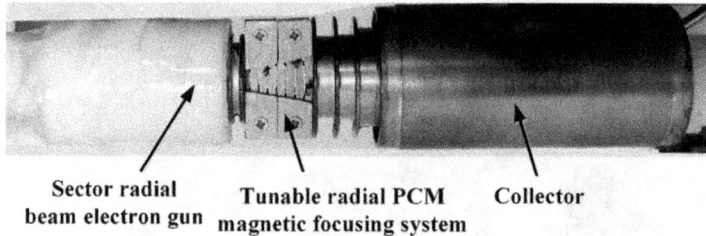

Figure 5.25. Flow tube with magnetic system installed. © [2017] IEEE. Reprinted, with permission, from [4].

circuits and high-voltage flaring between the electrodes of the electron gun. This effectively isolates the high voltage and protects the operators.

5.1.4 Overall assembly

After the transmission characteristics of the angular log-period meander strip-line SWS were tested in order to verify their reasonableness, the overall tube assembly of the angular log-period meander strip-line TWT was carried out. The TWT consists of four parts: the radially diverging zonal electron-beam gun, the angular log-period meander strip-line SWS, the collector, and the radially tunable PCM focusing system.

The electron gun, collector, exhaust tube, and assembled slow-wave system were welded in alignment, and the alignment accuracy was ensured through the use of fixtures. Laser welding and argon arc-welding processes were used to ensure the airtightness of the TWT.

The angular log-period meander strip-line TWT after integration and welding is shown in figure 5.26. The purpose of the evacuation process is mainly to form a high vacuum inside the TWT, which plays a key role in the working of the electron gun and cathode; it can also reduce the probability of ignition in the tube. After the welding and evacuation processes of the angular log-period meander strip-line TWT

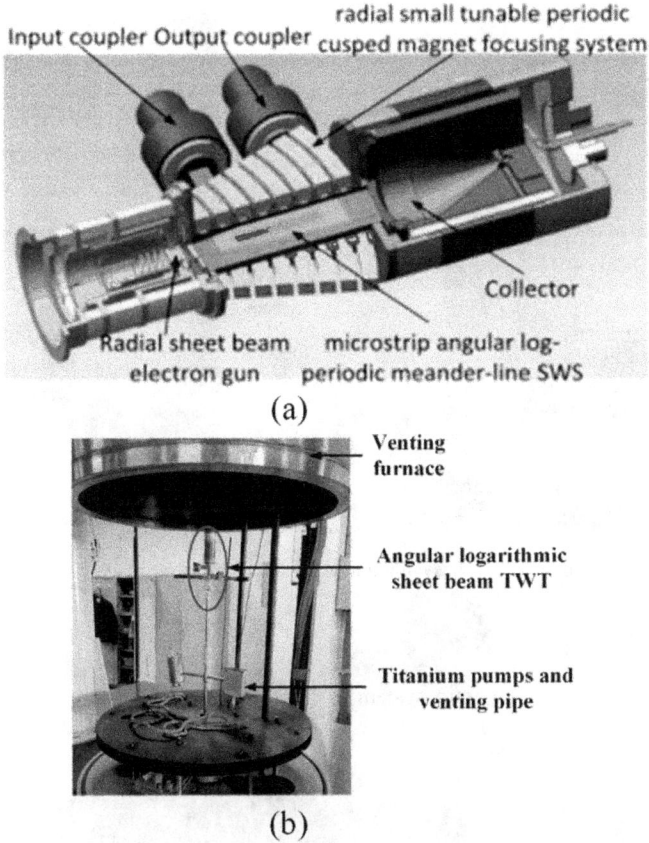

Figure 5.26. (a) Design schematic and (b) a photo of the assembled angular logarithmic strip TWT. © [2017] IEEE. Reprinted, with permission, from [4].

are complete, the assembly of the magnetic focusing system is carried out. Figure 5.27 shows the angular log-banded TWT after the assembly of the magnetic focusing system.

5.2 Modified angular log-period folded waveguide TWT

The overall structural model of the designed K_a-band MALP folded waveguide TWT is shown in figure 5.28. It has five main parts: the electron gun, the intermediate section, the collector pole, the box window, and the PCM field. The structural design of each part is introduced separately below.

5.2.1 Electron gun

Figure 5.29 shows the component composition and assembly structure of the designed Pierce electron gun; the parts represented by the labels and the materials used for their manufacture are listed in table 5.1.

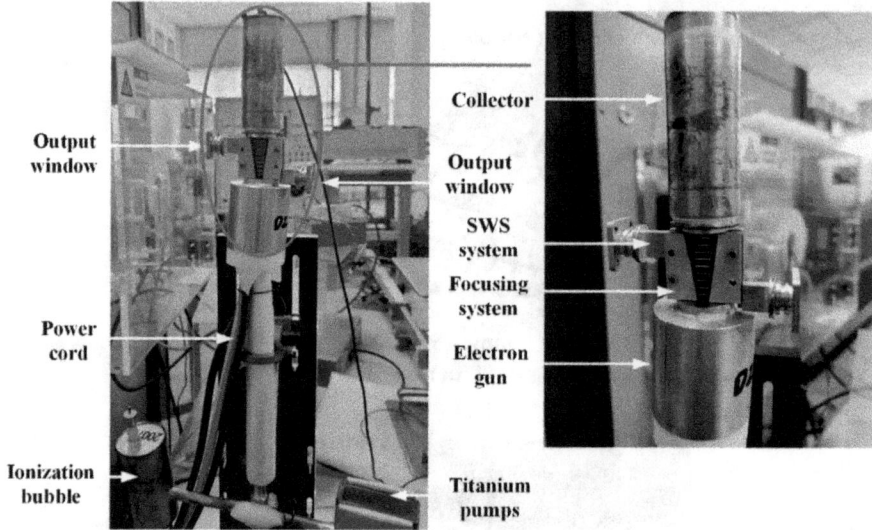

Figure 5.27. Assembled angular logarithmic strip TWTs.

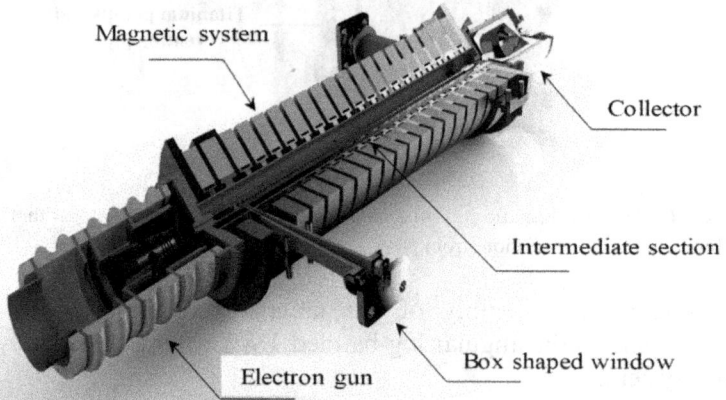

Figure 5.28. MALP folded waveguide TWT model intended for use in the K_a band.

It should be noted that behind the cathode there is actually a filament surrounded by ceramic, which is used to heat the cathode emitter, but this is not shown in figure 5.30.

The assembly of the electron gun is mostly done by laser welding. After a series of components is ready, they are positioned in turn on the fixture and then laser welded together, which is described in detail in this section.

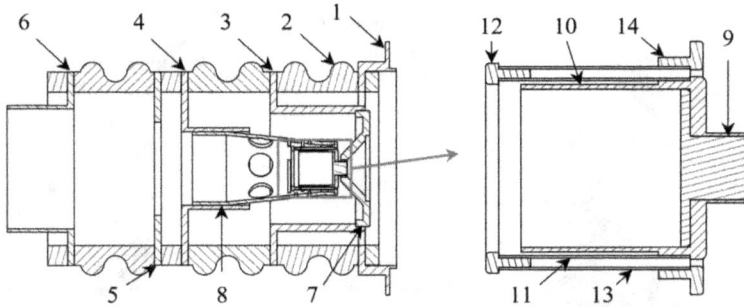

Figure 5.29. Structure of the electron gun.

Table 5.1. Part names and materials shown in figure 5.29.

Label no.	Name	Material	Label no.	Name	Material
1	Anode sealing ring	Kovar (4J34)	8	Support cylinder	Kovar (4J34)
2	Gun shell ceramic	Ceramic (A95)	9	Cathode emitter	Customized
3	Beam forming electrode (BFE) sealing ring	Kovar (4J34)	10	Cathode cylinder	Molybdenum
4	Cathode sealing ring	Kovar (4J34)	11	Internal heat shielding	Molybdenum–rhenium alloy
5	Filament sealing ring	Kovar (4J34)	12	Hot screen cylinder connection ring	Molybdenum
6	Titanium pump sealing ring	Kovar (4J34)	13	External heat shielding	Molybdenum–rhenium alloy
7	BFE	Molybdenum	14	External heat shielding clip	Molybdenum–rhenium alloy

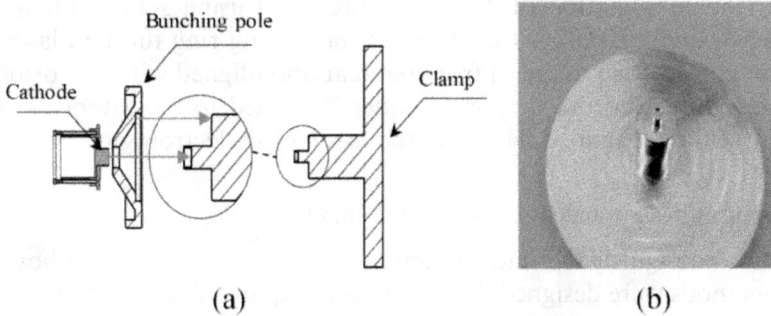

Figure 5.30. Clamp for positioning the gun. (a) Schematic of the assembly structure and (b) photograph of the sample.

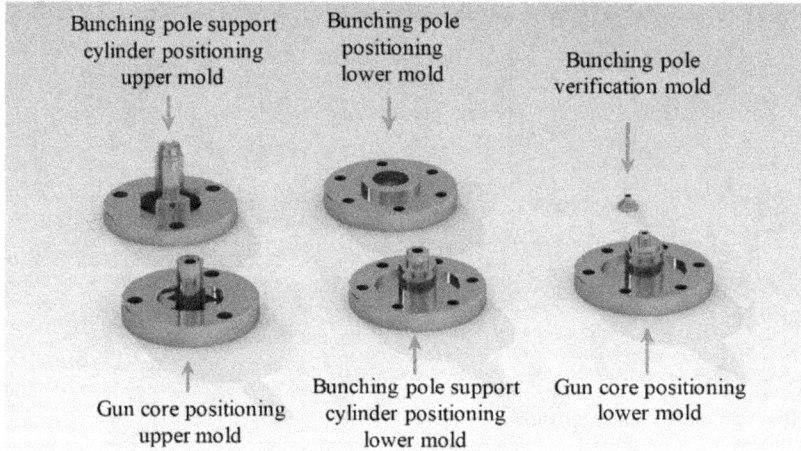

Figure 5.31. Photos of the gun fixtures.

The assembly sequence of each component is as follows: the beam-forming electrode support barrel, the gun core, the beam-forming electrode, the first anode, the second anode, and the back cover. The assembly requires fixtures to assist positioning, as shown in figure 5.31.

It can be seen that one end of the beam-forming electrode support cylinder is passed through the lower mold to ensure concentricity between it and the base-level, and the top end of the upper mold is held against the other end of the beam-forming electrode support cylinder. The distance and parallelism between the lower end face and the datum plane can be controlled by tightening and adjusting the screw. After the relative positions of the beam-forming electrode support cylinder and the gun shell are fixed, a laser beam is directed from above to weld the upper end face of the bunching pole support cylinder to the bunching pole sealing ring, shown in figure 5.32(a). After the fixture is removed, the welding position is as shown in figure 5.32(b), viewed from the rear end of the electron gun.

Figure 5.33 shows the assembly structure of the gun core and buncher.

The assembly sequence of the electron gun involves welding the gun core from the inside out (the combination of 8–14 in figure 5.29) using spot welding and laser welding. The bundling pole is then fixed to the sealing ring through laser welding. Finally, the gun core is pushed in from the rear and aligned with its position on the positioning fixture, and the support cylinder is welded to the cathode sealing ring using laser welding. Figure 5.34 shows the assembled electron gun.

5.2.2 Structural design and assembly of the middle section

The structural design of the middle section is shown in figure 5.35; however, two assembly methods were designed, both of which required fewer partial inter-rings to be installed during the brazing stage.

In figure 5.35, four inter-rings (A, B, C, and D) are marked. The first assembly method is similar to that used for the flow tube. It is a brazing-based approach. After

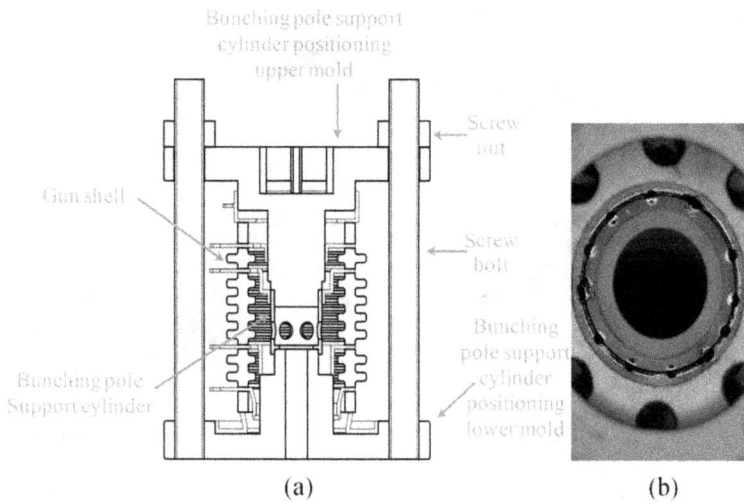

Figure 5.32. Assembly of the beam-forming electrode support cylinder. (a) Assembly structure and (b) post-welding photo.

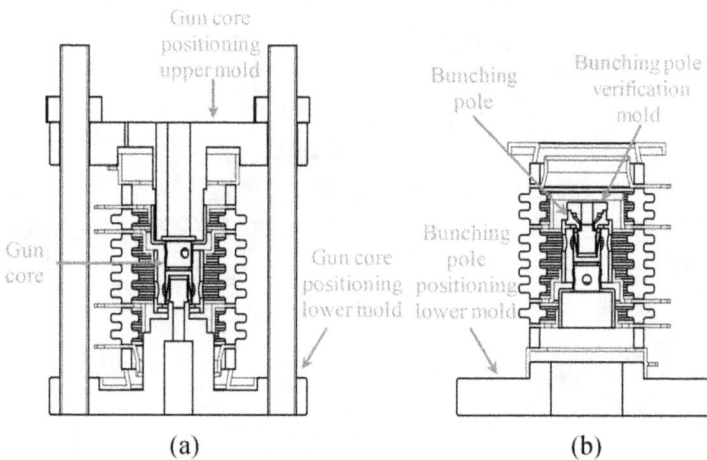

Figure 5.33. Assembly structure of the electron gun. (a) Gun core and (b) beam-forming electrode.

brazing, all four inter-rings need to be removed, and all magnetic screens, inter-rings, and pole boots are then movable. The advantages of this assembly method are that there are fewer welds and it is easy to meet the requirement for airtightness. Figure 5.36 shows a photo of the first middle segment. The second assembly method uses two brazes. The first brazes the interface, magnetic screen, pole shoe, and inter-ring into three composite components, and the second brazes these components together with the SWS, sleeve, input/output waveguide, and waveguide welding edges to form the intermediate segment, as shown in figure 5.37.

In order to ensure the concentricity of the various parts during the welding of the composite components, they need to be assembled on a fixture with a core rod to

(a)   (b)

Figure 5.34. Photo of the electron gun. (a) General perspective and (b) cathode surface.

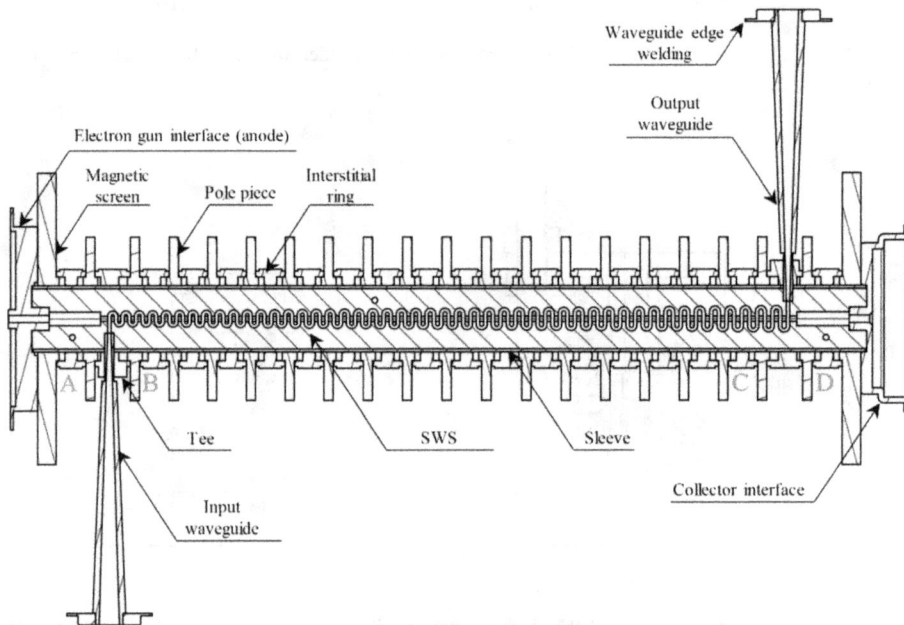

Figure 5.35. Structural design and assembly of the middle section.

ensure that the SWS and the sleeve can be smoothly inserted into the second brazing process. Figure 5.38 shows a photo of the welded second middle section.

The two assembly methods have advantages and disadvantages: the first assembly method has simple procedures, fewer welds, and makes it easy to achieve airtightness, but its disadvantages are the same as those of the flow tube; that is, the mobile pole piece and the inter-ring are not conducive to debugging the electron-beam flow. The second assembly is exactly the opposite of the first: the pole piece, ring, and other parts are firmly fixed to the sleeve, but the process is relatively more complex; not only does it need two brazes but also a lot of welds (there are 46 welds). In

(a)   (b)

Figure 5.36. Middle segment. (a) Waiting for welding to take place in the hydrogen furnace and (b) after completion of the welding.

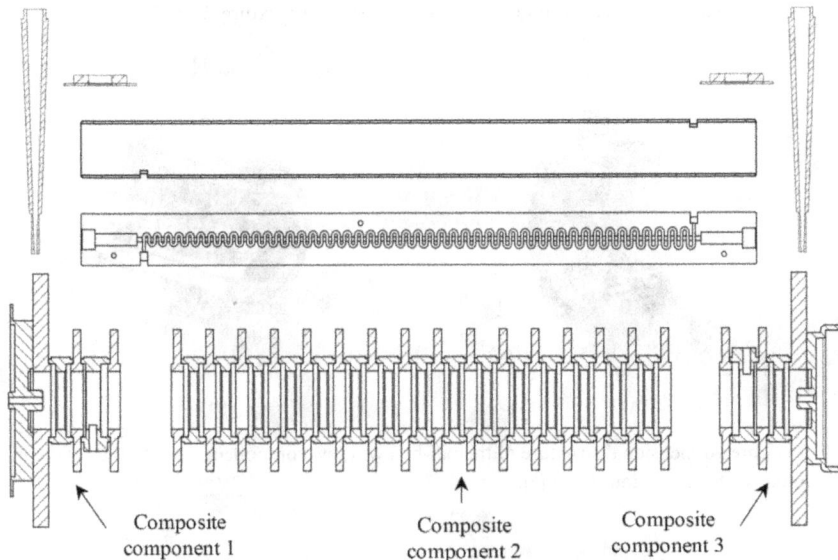

Composite component 1   Composite component 2   Composite component 3

Figure 5.37. Components of the second assembly method.

addition, its composite components not only need to ensure airtightness but also control cumulative errors.

5.2.3 Other components

The structures of the box window and the collector are relatively simple. Figure 5.39 shows their models.

As can be seen from figure 5.39, the structure of the pill-box window is very simple, including only three parts, i.e. the upper and lower window frames and the window piece. The metallized window piece is sandwiched between the upper and lower window frames, and a circle of solder wire is placed around the edge of the

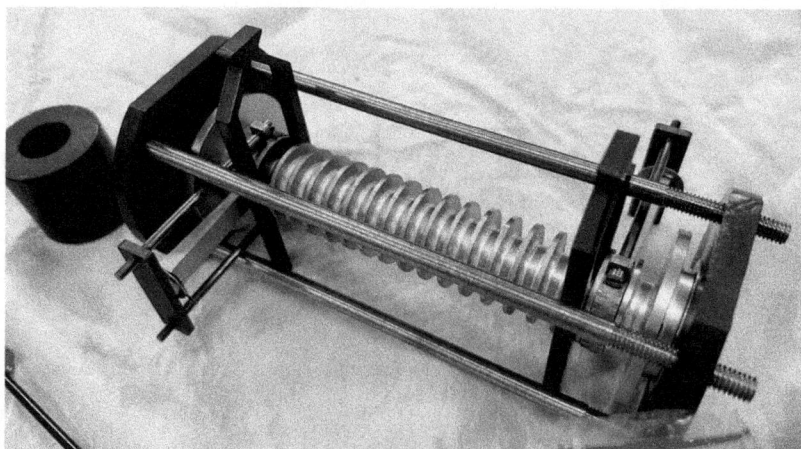

Figure 5.38. Second middle section placed on the fixture for welding.

(a)   (b)

Figure 5.39. (a) Three-dimensional structure of the pill-box window component, and (b) the collector. © [2023] IEEE. Reprinted, with permission, from [5].

window to complete its welding. The material of the window is sapphire, and the relative dielectric constant used in the simulation design is 9.4. The window frame is made of grade 4J34 nickel-plated Kovar alloy.

Figure 5.40 shows the four-stage depressed collector components, which include four electrodes, the rear sealing ring, the front sealing ring, the shell, the collector bottom, the rear connection, and several ceramics. The outer conductor, the inner conductor, and the insulating ceramic are constructed as follows: the outer conductor comprises a shell and a sealing ring, the inner conductor comprises an inner cylinder and a lead, and the insulating ceramics support the inner cylinder and the lead. Figure 5.41 shows the assembled collector. The airtightness and insulation tests of this design meet the requirements.

The assembly of the magnetic ring draws on the experience gained from the previous flow tube. First of all, mechanical cutting rather than manual cutting is used to cut the magnetic ring, and custom-sized clasp rings are also used to fix the

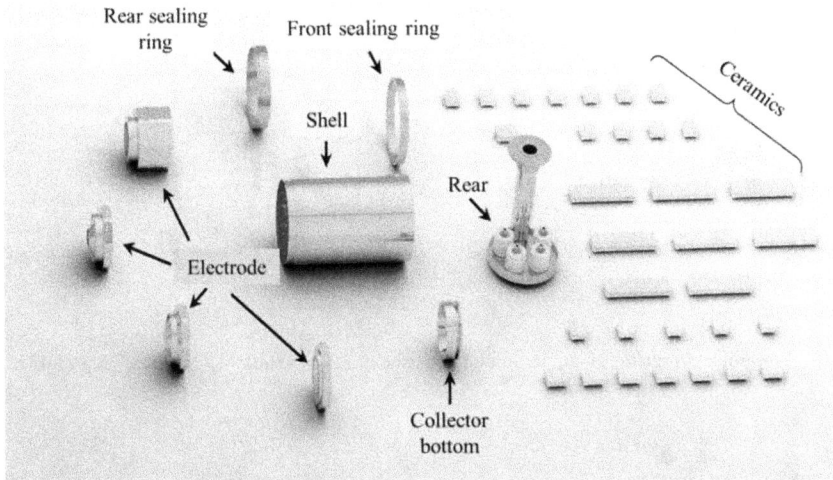

Figure 5.40. Components of a four-stage depressed collector.

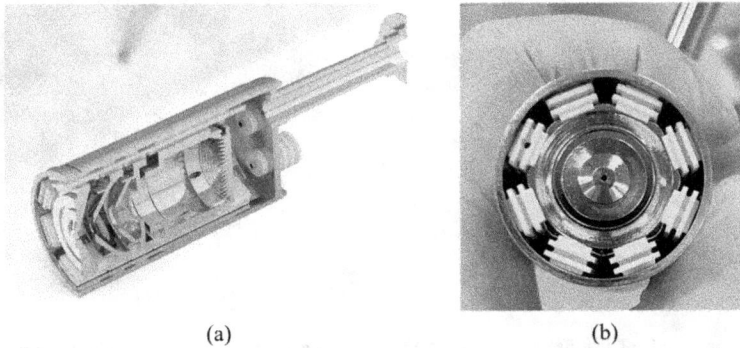

(a)　　　　　　　　　　　　　　(b)

Figure 5.41. Assembled collector. (a) Model and (b) frontal photographs.

magnetic ring. The assembled structure consisting of the magnetic ring and the clasp ring is shown in figure 5.42.

The connections between the electrode and the ceramic, between the electrode and the lead, between the ceramic and the housing, and between the housing and the sealing ring are all completed by brazing. The rear sealing ring and the rear cover are welded by continuous laser welding.

5.2.4 Overall TWT assembly

Once the individual components are ready, the assembly of the whole tube is relatively simple. The argon arc-welding process is used for the welding, which requires a total of seven welds, as follows: between the electron gun and the middle section, between the collector and the middle section, between the two windows and the middle section, the connections between the electron gun back cover and the two

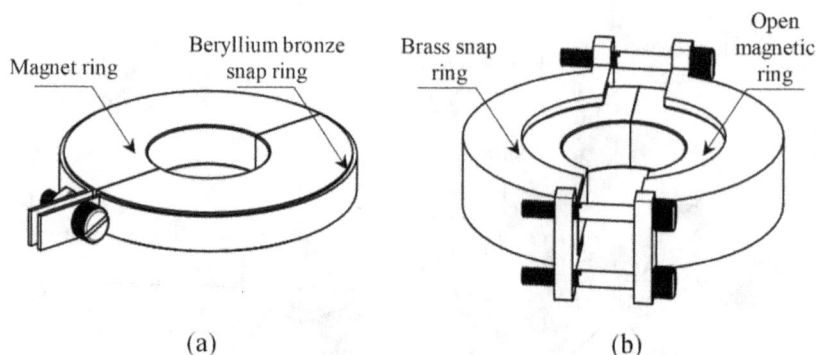

(a)   (b)

Figure 5.42. Assembled structure of the magnetic ring and the clasp ring. (a) A whole ring and (b) an open ring.

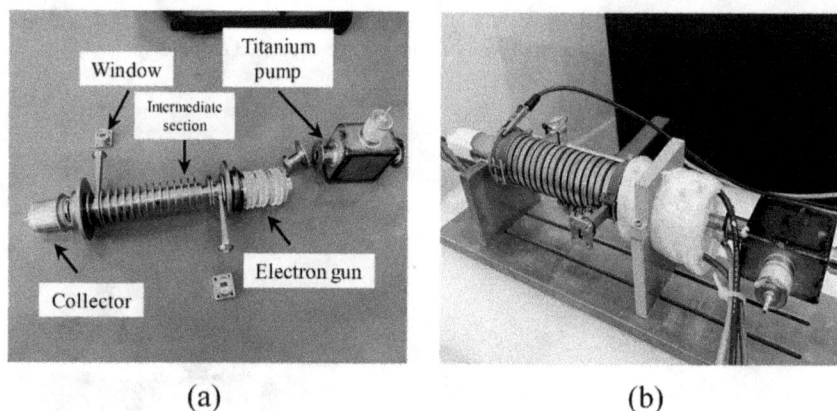

(a)   (b)

Figure 5.43. Photograph of the first K_a-band MALP folded waveguide TWT. (a) Components and (b) cold high-voltage test. © [2023] IEEE. Reprinted, with permission, from [5].

titanium pumps, and between the titanium pump and the row tube. Figure 5.43(a) shows the main components of the first K_a-band MALP folded waveguide TWT. After the welding is complete, it is also necessary to evacuate all gas, connect the cables, and glue and install the magnets. Figure 5.43(b) shows the first K_a-band MALP folded waveguide TWT.

References

[1] Wang S, Gong Y, Hou Y, Wang Z, Wei Y, Duan Z and Cai J 2013 Study of a log-periodic slow wave structure for Ka-band radial sheet beam traveling wave tube *IEEE Trans. Plasma Sci.* **41** 2277–82

[2] Li X, Xu Y, Wang S, Wang Z, Shi X, Duan Z, Wei Y, Feng J and Gong Y 2016 Study on phase velocity tapered microstrip angular log-periodic meander line travelling wave tube *IET Microw. Antennas Propag.* **10** 902–7

[3] Wang W 2014 *Microwave Engineering Technology* 2nd edn (National Defense Industry Press)

[4] Li X, Wang Z, He T, Gong H, Duan Z, Wei Y and Gong Y 2017 Study on radial sheet beam electron optical system for miniature low-voltage traveling-wave tube *IEEE Trans. Electron Devices* **64** 3405–12

[5] Xu D, Wang S, Lu C, He T, Wang Z, Lu Z, Gong H, Duan Z and Gong Y 2023 Demonstration of a modified angular log-periodic folded waveguide traveling wave tube at Ka-band *IEEE Trans. Electron Devices* **70** 1323–9

Chapter 6

System tests

The experimental testing mainly consists of component testing and whole-tube testing [1, 2]. Component testing includes measuring the characteristic size of the slow-wave structure, measuring the magnetization of the magnetic ring, and testing the transmission characteristics of components related to the middle segment [3–5].

6.1 Angular radial log-periodic meander-line traveling-wave tube

6.1.1 Beam flow test

Figure 6.1 shows the flow tube test platform connected to the test system. The test system consists of a high-voltage power module and an oscilloscope. The electron gun of the flow tube ensures that no electrode discharge occurs during high-voltage operation after gluing and vacuuming. The part wrapped in polytetrafluoroethylene (PTFE) film is the exhaust tube, which is at the same potential as the cathode. The PTFE film isolates the high voltage and protects the operator.

The test voltage of the flow tube is 1700 V. The heating of the cathode heater takes 3 min according to the operating procedures. During this process, the current of the hot wire slowly rises to 1 A, reaching the temperature required for the normal operation of the cathode. The cutoff voltage of the beam-forming electrode and the cathode voltage are added in turn. The cutoff voltage of the beam-forming electrode is turned off at the start of debugging. At this time, the cathode is in the emission state.

In the debugging process, the main adjustment part is the pole piece. Using a pointed, tough, non-magnetic tool, the pole piece is gently moved along the guide groove on the fixture; its position is secured, and the next piece of the pole piece is adjusted when the flow reaches the optimum value that the pole piece can achieve. Because the position of the sector magnetic block is completely restricted by the fixture and cannot be moved, the compensation of the magnetic block is mainly achieved by using the pole piece to find a better flow rate state at a given position of the magnetic block.

doi:10.1088/978-0-7503-5452-3ch6   6-1

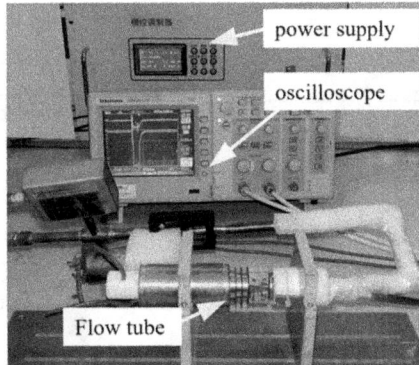

Figure 6.1. Flow rate test system.

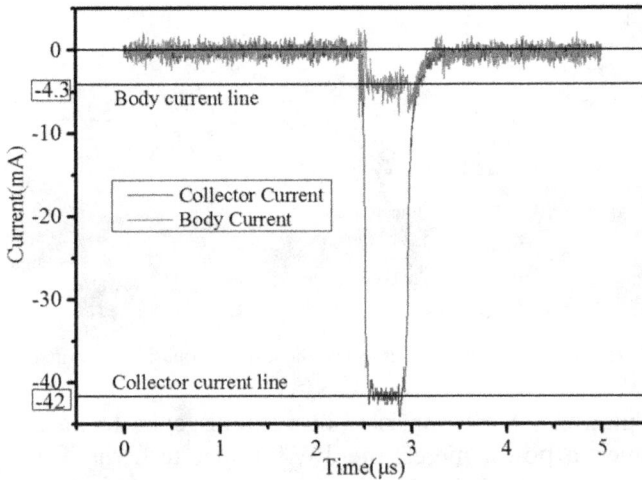

Figure 6.2. Flow rate data curve obtained from the oscilloscope. © [2017] IEEE. Reprinted, with permission, from [6].

The current measurements derived from the oscilloscope are shown in figure 6.2. The signal collected by the oscilloscope is the voltage signal when the current flows through the resistance. According to the relationship between voltage, resistance, and current given by Ohm's law, the current value can be converted to obtain the relationship between the current of the tube and the current of the collector, respectively. As can be seen from the figure, the collector current is 42 mA, the tube body current is 4.3 mA, and its flow rate is 90.7%.

6.1.2 Overall tube test

After assembly was completed, experimental work was carried out to verify the reasonableness of the angular log-period meander-line traveling-wave tube (TWT) . Figure 6.3 shows the thermal test system framework used to test the angular log-strip

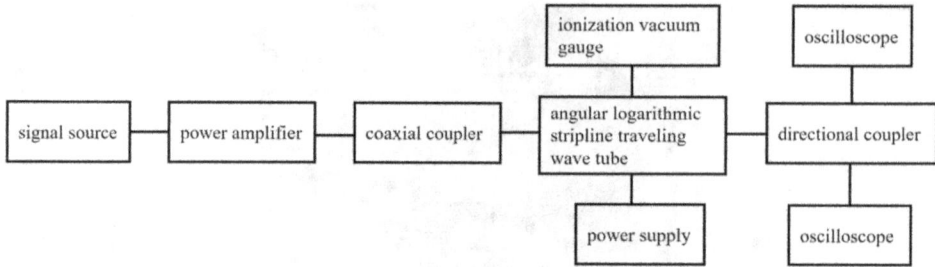

Figure 6.3. Framework of the thermal test system.

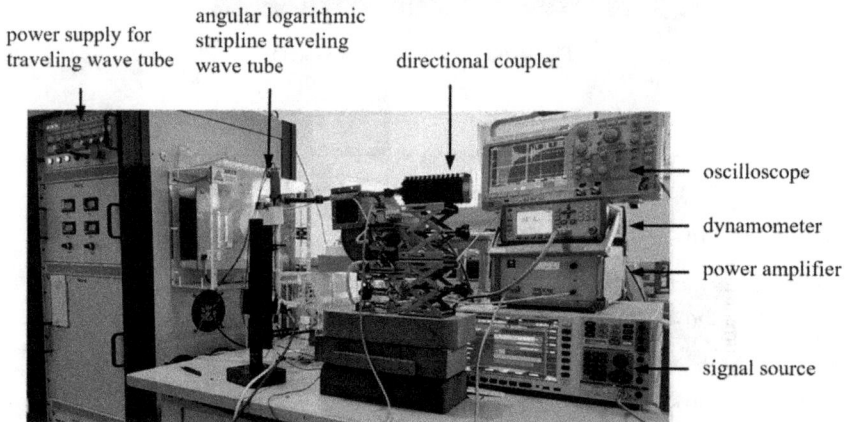

Figure 6.4. Test platform used to test the angular logarithmic strip TWT.

TWT. The test platform TWT consisted of a signal generator, a power amplifier, a directional coupler, a power meter, the TWT under test, the TWT power supply, various transmission conversion cables, and adapter connectors, as shown in figure 6.4.

The working voltage of the TWT electron gun was 5700 V, the duty cycle was 0.1%, and the pulse repetition frequency was 5 Hz. The cathode filament voltage was 6.47 V, and the filament current was 1 A. By adjusting and supplementing the focusing magnetic field, the electron beam flow rate of the TWT was adjusted to maximize the electron beam flow rate. After debugging, the current data for the tube body and the collector shown in figure 6.5 confirmed that the emission current of the electron gun was 52 mA, the measured current of the collector was 36 mA, and the flow rate of the electron beam was about 69%.

After the electron beam reached a stable flow state, the power test of the TWT was carried out. Figure 6.6 shows a comparison between the simulation results of the angular logarithmic strip slow-wave structure and the experimental results of thermal measurement under working conditions of a 50 mA current and an 80 mW input signal. A gain of more than 2 dB was obtained in the range of 32.6 GHz to 36.2 GHz, and the maximum gain of 4.4 dB was achieved at 34 GHz.

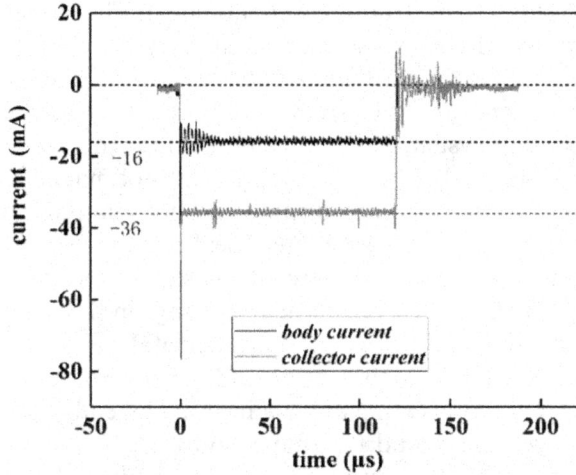

Figure 6.5. Tube body current and collector current.

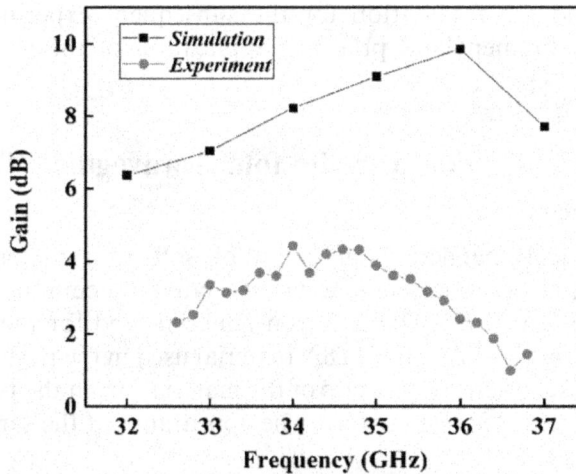

Figure 6.6. Comparison between experimental test results and simulated results.

As can be seen from the figure, for a small signal input of 80 mW, the gain of the simulation results was 6–10 dB in the range of 32–37 GHz, and the gain of the experimental test was 2–4.4 dB. There was a 4–6 dB gap between the two, and the measured output power working band was shifted to the low-frequency end, but the overall trend was in good agreement.

The difference between the hot measurement results and the simulation results was mainly caused by the low electron beam flow rate. The low flow rate of the electron beam was caused by many factors. On the one hand, the assembly accuracy and alignment accuracy are poor: the assembly error could not be accurately controlled during the assembly of the electron gun parts and the assembly of the electron gun and slow-wave structure parts, and the parallelism and shape control of the cathode and the slow-wave structure were not ideal. The low alignment accuracy of the electron gun and the slow-wave structure may have led to a change in the emission angle and the height of the electron beam and affected the shape of the electron beam when entering the slow-wave structure. On the other hand, the design of the magnetic focusing system had a certain ideality. In the actual test, the flow rate of the electron beam had to be adjusted by means of magnet compensation and adjustment. The shape of the electron beam could not maintain an ideal shape and was quite different from that of the simulated ideal electron beam. At the same time, some deformation in the assembly process of the slow-wave structure also led to a change in the overall structural fit, resulting in a change in the working voltage of the slow-wave structure, thus affecting the experimental results. Therefore, in the subsequent experimental work, optimization and improvement of the scheme should be carried out in the parts processing and engineering assembly so as to improve the flow rate of the electron beam and the shape control.

The gain amplification of the output signal was obtained through a sample tube thermal measurement experiment of the angular log-period meander-line TWT. The reasonableness of the slow-wave structure of the angular log-strip TWT was verified, which laid the foundation for the subsequent experimental work and promoted the development and practical application of the miniaturized radial beam TWT.

6.2 Modified angular log-periodic folded waveguide TWT

6.2.1 Magnetic circuit test

Although the typical magnetization curve of soft iron materials is given in simulation software, the materials used in actual machining are not always consistent with this curve [8]. The soft iron material used for the magnetic screen and the pole piece was DT8A. The DT8A material used in the parts processing came from two different suppliers, so samples of the materials from the two suppliers were taken, labeled AA and BB, respectively. The appearance of the samples is that of Φ 9.9 mm $\times$ 100 mm rod.

The test instrument was a C-750 static hysteresis loop tester made by Magnet-Physik, Germany. The test method was GB/T13012-2008. The ambient temperature was 25 °C, and the relative humidity was 60%. The test results of the two samples are shown in figure 6.7.

The test results in figure 6.7 show that the magnetization curve of the AA sample is higher than that of the BB sample and is closer to the typical curve given in the Opera software. Therefore, the magnetically conductive parts such as the magnetic screen and pole boots used in the subsequent experiments were all made of DT8A soft iron materials provided by the supplier of the AA sample, and the test results of the AA sample were imported into the simulation software in the subsequent magnetic system simulation instead of using the default typical curve.

The measurement of the bare magnetic ring includes the measurement of the size and the magnetic induction intensity of the axis (figure 6.8).

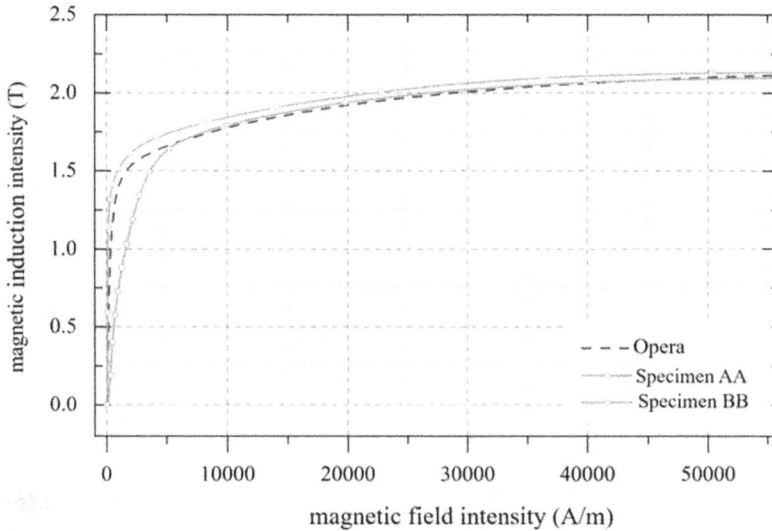

Figure 6.7. Test results for the magnetization curves of the DT8A soft iron material samples. © [2022] IEEE. Reprinted, with permission, from [8].

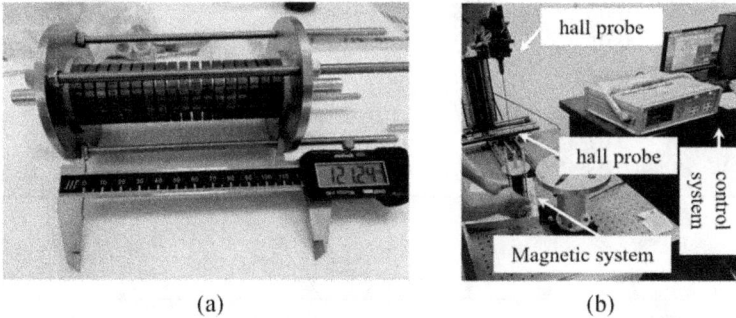

Figure 6.8. Photo of a magnetic system test. (a) Magnetic system mounted on a fixture; (b) test platform. © [2022] IEEE. Reprinted, with permission, from [8].

The two sets of magnetic field test results are shown in table 6.1. The data shows that the magnetic induction intensity of the axis of the bare magnetic ring is consistent with the design value, the maximum error is −0.0006 T (which appears in the 14th magnetic ring of the second set of magnetic rings), and the error control of the first set of magnetic rings is slightly better than that of the second set. Therefore, the first set of magnetic rings was used in the flow tube experiment assembled in this section.

The test results show that the magnetic induction intensity curve of the axis of the system is very consistent with the simulation results, and the overall strength is slightly higher than the simulation results (figure 6.9). This is because the magnetization curve test results of soft iron materials were not used in the design; the typical

Table 6.1. The test results for the dimensions of the bare magnetic ring and the magnetic induction strength of the axis.

| | Dimension values | | | | B_z | | |
| | 1st set | | 2nd set | | | | |
No.	Outer diameter (40 mm)	Thickness (5.2 mm)	Outer diameter (40 mm)	Thickness (5.2 mm)	Design value (T)	1st set (T)	2nd set (T)
1	0.05	0.05	0.05	0.05	0.0668	0.0666	0.0668
2	0.05	0.05	0.03	0.06	0.1224	0.1226	0.1227
3	0.03	0.04	0.03	0.04	0.1510	0.1508	0.1511
4	0.03	0.06	0.02	0.06	0.1589	0.1588	0.1591
5	0.02	0.07	0.04	0.04	0.1589	0.1589	0.1587
6	0.05	0.06	0.02	0.05	0.1589	0.1591	0.1590
7	0.03	0.05	0.05	0.05	0.1589	0.1587	0.1586
8	0.03	0.06	0.05	0.05	0.1589	0.1585	0.1591
9	0.04	0.06	0.05	0.04	0.1589	0.1592	0.1586
10	0.03	0.04	0.02	0.05	0.1589	0.1589	0.1585
11	0.02	0.06	0.05	0.05	0.1589	0.1588	0.1587
12	0.03	0.07	0.03	0.04	0.1589	0.1590	0.1591
13	0.05	0.06	0.04	0.04	0.1589	0.1590	0.1591
14	0.05	0.05	0.03	0.05	0.1589	0.1588	0.1583
15	0.04	0.05	0.05	0.06	0.1558	0.1558	0.1559
16	0.03	0.06	0.02	0.07	0.1558	0.1560	0.1556
17	0.03	0.07	0.03	0.07	0.1303	0.1307	0.1302
18	0.03	0.05	0.04	0.06	0.1001	0.1002	0.0999

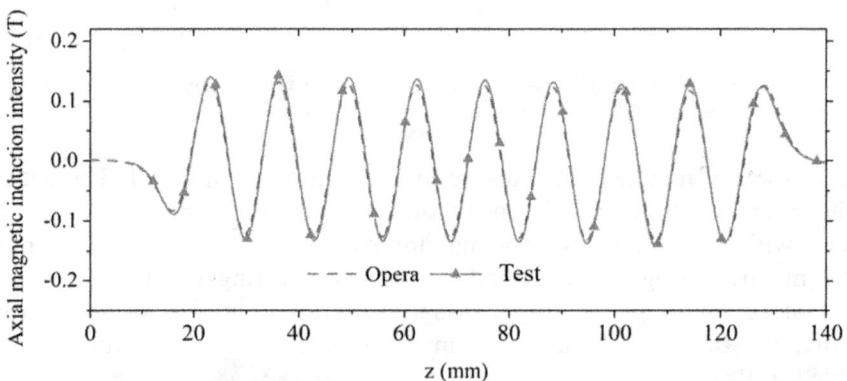

Figure 6.9. Test results for the axial magnetic induction intensity. © [2022] IEEE. Reprinted, with permission, from [8].

curve given by Opera was used, while the magnetization curve of the actual soft iron materials was slightly higher than the typical curve.

The overall structure of the flow tube is shown in figure 6.10. It includes four main parts: the electron gun, the intermediate segment, the magnetic ring, and the collector electrode. Photos of the parts are shown in figure 6.11. The electron gun, middle section, and collector electrode are connected by argon arc welding.

There are only two welds in the design of the structure, as shown in figure 6.12. The airtightness of the structure is mainly guaranteed by the sleeve and the two welds. This design compensates for leaks by moving these moving parts if the welding fails to achieve airtightness. The magnetic ring is split before installation, and the inter-ring that is not installed during brazing is also split. The split parts are finally installed together in pairs on the middle section and tightened by the clasp ring and copper wire. Figure 6.13 shows a photo of the assembled flow tube.

Table 6.2 lists the temperature test data. The middle section is welded using a hydrogen furnace brazing process, and the solder is $Ag_{72}Cu_{28}$. In figure 6.11(b), it can be seen that the middle section is brazed without directly mounting all the inter-rings at once and that the pole shoe, inter-ring, and magnetic screen are not directly welded to the sleeve by brazing filler metal.

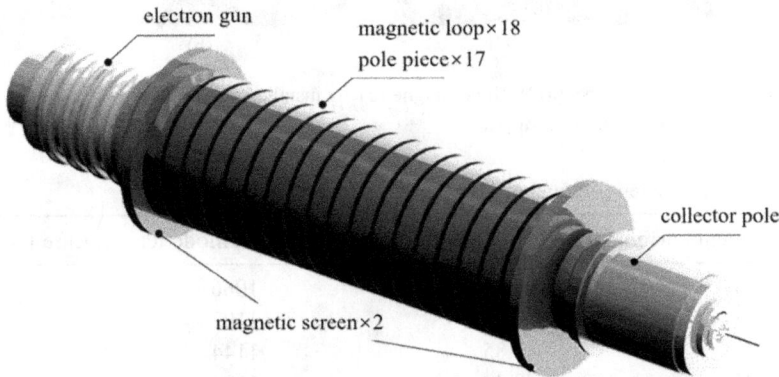

Figure 6.10. Overall structure of the flow tube. © [2022] IEEE. Reprinted, with permission, from [8].

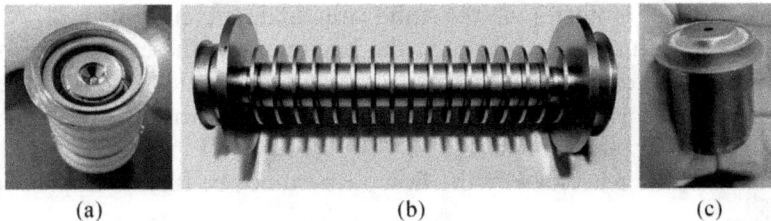

Figure 6.11. Photos of the components of a beam flow tube. (a) Electron gun; (b) intermediate segment; (c) collector. © [2022] IEEE. Reprinted, with permission, from [8].

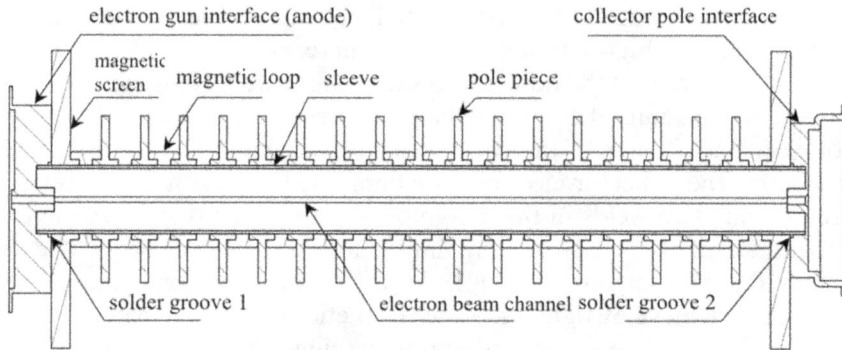

Figure 6.12. Structure of the intermediate segment.

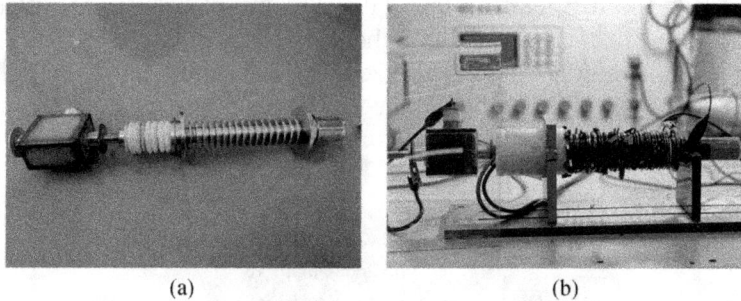

Figure 6.13. Photos of a flow tube. (a) With no magnetic ring installed and no wiring; (b) under test. © [2022] IEEE. Reprinted, with permission, from [8].

Table 6.2. The temperature test data of flow tube electron gun.

Filament current (A)	Filament voltage (V)	Cathode temperature (°C)
1.655	6.86	1066
1.724	7.28	1102
1.821	7.85	1144
1.925	8.49	1194

The power-on parameters of the tube are: filament current 1.74 A, cathode voltage −12.1 kV, duty cycle 0.1%. At the beginning of the cathode emission, the current is 96 mA; after a week of aging, the cathode emission current is reduced to 93.9 mA; after debugging, the final measured collector current is 82.9 mA and the corresponding flow rate is about 88.3%.

After analysis, the reasons why the flow rate cannot be further improved may be as follows:

(1) The filling ratio of the electronic beam in the design is very high, and the filling ratio of the line at the maximum envelope radius in the simulation

exceeds 93%. The electronic beam is very close to the edge of the channel, so the flow rate is sensitive to assembly errors.

(2) The general debugging process requires repeated rotation of the magnetic ring, but because the pole piece, inter-ring, and other parts in the design of this section are active, the rotation of a single magnetic ring is associated with the concurrent rotation of several nearby magnetic rings, so that independent control of the rotational angle of a single magnetic ring cannot be achieved.

(3) The typical magnetization curve of the soft iron material used in the design is different from the experimental test results, resulting in a slightly larger magnetic field than the design value.

(4) The size of the magnetic ring used in this experiment is large, and the installation is manually divided, so the profile is uneven. In addition, the clasp is not a customized part, and the size and magnetic ring are not well matched, which leads to large gaps in the magnetic ring assembly.

Subsequent improvement methods to address the above deficiencies include:

(1) Reduce the filling ratio of electronic injection in the design, leaving a larger design margin.

(2) Change the structural design and welding method of the middle section, fixing the pole shoe, inter-ring, and other parts to avoid rotation.

(3) When designing the magnetic field, use the experimentally measured magnetization curve to replace the typical magnetization curve provided in the software.

(4) When the magnetic ring is split, mechanical machining should be used to ensure that its profile is smooth, and the clasp ring should be customized according to the size of the magnetic ring so as to minimize the dislocation problem in the magnetic ring assembly.

6.2.2 Overall tube test

Figure 6.14 shows the structure and photos of the TWT hot measurement platform. The sampling attenuations of power meter 1 and power meter 2 are 19.7 dB and 32 dB, respectively. The input port of the spectrometer is not fixed to the sampling output port of the directional coupler 3 but is only placed nearby to prevent it from being damaged by phenomena such as tube ignition. The signal source continues to output the signal, but the same pulse signal is added to the cathode and the beam-forming electrode.

Figure 6.15 shows the static flow rate test results under different working voltages, debugging has taken place after magnetic field compensation.

The experimental results show that the emission current of the cathode increases with an increase in the working voltage. When the working voltage is in the range of 12–12.8 kV, the emission current of the cathode is 70.2–76.3 mA, which is obviously lower than the design value of 100 mA. Under these conditions, the measured electron flow rate in this voltage range is 83%–87.2%.

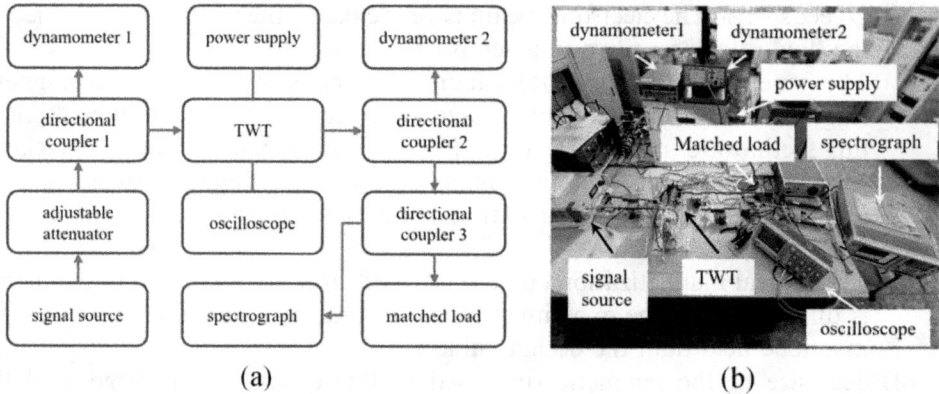

Figure 6.14. Platform used to test the modified angular log-period (MALP) folded waveguide TWT in the Ka band. (a) Structure; (b) photographs. Part (b) © [2023] IEEE. Reprinted, with permission, from [7].

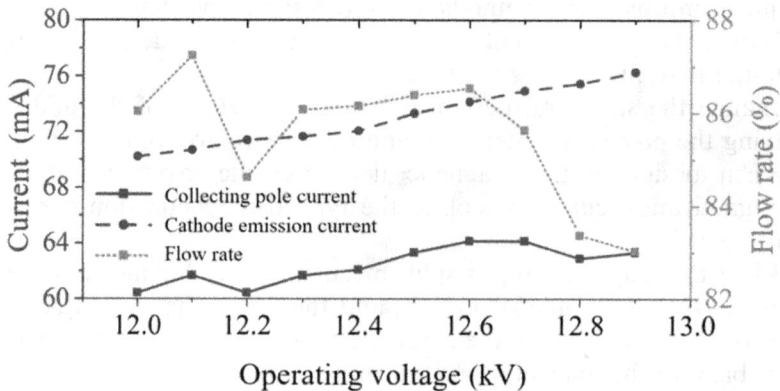

Figure 6.15. Static flow rate test results for the MALP folded waveguide TWT in the Ka band. © [2023] IEEE. Reprinted, with permission, from [7].

Figure 6.16 shows the final hot measurement results of the TWT, where the input/output waveguide and window losses have been substituted for the corrected results because they were not included in the model used to simulate the time and accuracy of the beam–wave interaction. In the experiment, the input power of each frequency point is fixed at 8.9 mW, corresponding to −10 dBm displayed on the power meter 1, while the input power of the slow-wave structure's inlet corresponds to 7 mW and the working duty cycle is 0.45%. An additional set of beam–wave interaction simulation results is shown in figure 6.16, corresponding to a reduction of the operating current to 75 mA in the previous simulation model.

At a frequency of 32.5 GHz, the power displayed on power meter 2 is −26.57 dBm when the tube is unloaded, while the power displayed when the tube is loaded is −12.6 dBm. The pulse power at the position of the output flange of the TWT can be obtained by performing a conversion based on the duty cycle and sampling

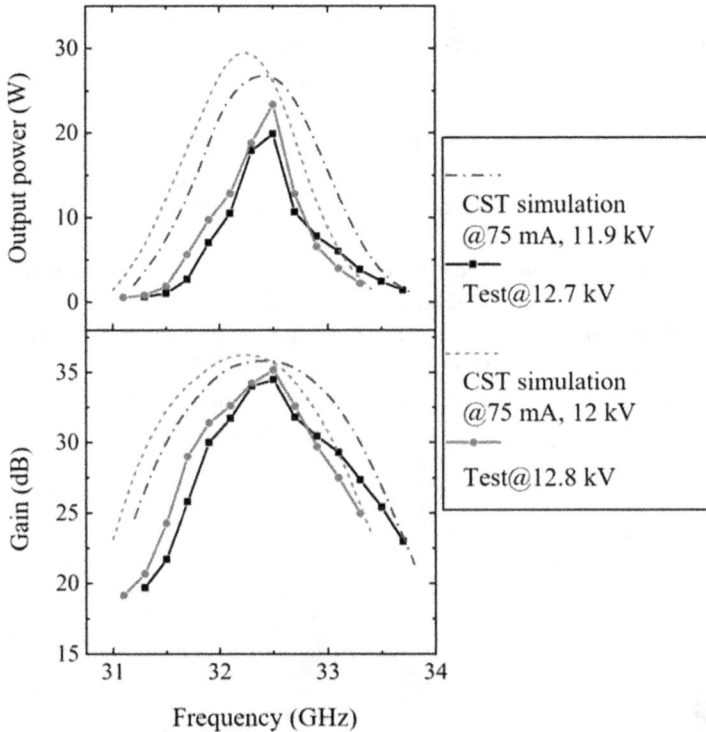

Figure 6.16. Thermal measurement results for the MALP folded waveguide TWT in the Ka band. © [2023] IEEE. Reprinted, with permission, from [7].

attenuation data. The pulse power at the position of the output flange is found to be 18.58 W. The pulse power at the output port position of the slow-wave structure is 23.39 W. Other frequencies are also converted in the same way.

A comparison between the test results and the simulation results shows that although the operating voltage difference between the two is 800 V, the overall trend of their output power and gain curves is consistent. However, there are still some differences in value. Taking the 12.8 kV experimental curve and the 12 kV simulation curve as an example, in the frequency range of 31.3–33.7 GHz, the simulated gain is 25.5–36.3 dB, while the experimental gain is 19.2–35.2 dB. The main reason for this may be that the motion state and trajectory of the electrons in the slow-wave structure may be different from those in the simulation after the emission current of the electron gun is reduced.

References

[1] Wang W 2014 *Microwave Engineering Technology* 2nd edn (Beijing: National Defense Industry Press) (in Chinese) pp 414–8

[2] Basu B N 1996 *Electromagnetic Theory and Applications in Beam-Wave Electronics* (Singapore: World Scientific) pp 381–94

[3] Booske J H, Converse M C, Kory C L, Chevalier C T, Gallagher D A, Kreischer K E, Heinen V O and Bhattacharjee S 2005 Accurate parametric modeling of folded waveguide circuits for millimeter-wave traveling wave tubes *IEEE Trans. Electron Devices* **52** 685–94

[4] Xu D *et al* 2020 Theory and experiment of high-gain modified angular log-periodic folded waveguide slow wave structure *IEEE Electron Device Lett.* **41** 1237–40

[5] Antonsen T M, Vlasov A N, Chernin D P, Chernyavskiy I A and Levush B 2013 Transmission line model for folded waveguide circuits *IEEE Trans. Electron Devices* **60** 2906–11

[6] Li X *et al* 2017 Study on radial sheet beam electron optical system for miniature low-voltage traveling-wave tube *IEEE Trans. Electron Devices* **64** 3405–12

[7] Xu D *et al* 2023 Demonstration of a modified angular log-periodic folded waveguide traveling wave tube at Ka-band *IEEE Trans. Electron Devices* **70** 1323–9

[8] Xu D *et al* 2022 Experiment of a high fill ratio electro-optical system for a Ka-band traveling wave tube *2022 23rd Int. Vacuum Electronics Conf. (IVEC) (Monterey, CA, USA)* pp 312–3

IOP Publishing

Planar Slow Wave Structure Traveling Wave Tubes
Design, fabrication and experiment
Yubin Gong and Shaomeng Wang

Chapter 7

Future perspectives and discussion

Even though planar slow-wave structures (SWSs) have attracted intensive attention because of their simple structure, easy access, low cost, high integration, etc. we have to admit that we still have a long way to go before we finally achieve success. Today's advanced fabrication methods are optimized for semiconductor devices, which are different from those of vacuum devices in terms of dimensions, materials, and environment. Therefore, the capability of advanced fabrication methods should be improved, and planar SWSs should be further revised [1–4].

7.1 Future direction

The employment of uniform periodic waveguides, such as the rectangular waveguide, the microstrip, the strip line, etc. is the simplest way to construct an SWS [5]. However, more and more evidence has proven that nonuniform periodic or quasi-periodic waveguides can provide much better performance, i.e. higher output power, higher electron efficiency, and so on. Especially in the millimeter-wave frequency band or above, as the structural dimensions become smaller, the beam–wave interaction length becomes a sensitive quantity, which means that a tiny change in the period leads to an obvious effect on the final result. This section introduces the techniques of hybrid beam–wave interaction.

7.1.1 Hybrid dispersion topology

It is known that every SWS has its own dispersion characteristics, which directly determine the performance of the device. Generally, an SWS with stronger dispersion characteristics (SD SWS) has a narrower operational frequency band and higher interaction efficiency than an SWS with weak dispersion characteristics (WD SWS). The concept of an SWS with hybrid dispersion (HD SWS) is realized by using two or more SWSs with different dispersion characteristics at the same operational frequency; these are alternately connected to form a single-section SWS [6]. Therefore, the HD SWS is characterized by balanced interaction efficiency

doi:10.1088/978-0-7503-5452-3ch7

over a moderate bandwidth. This concept can be applied to any type of SWS, and the comb SWS is taken as an example to introduce the technique in detail.

A comb SWS can be obtained by loading deep cavities or vanes periodically on the narrow edge of a rectangular waveguide. Its dispersion characteristics are determined by the depth of the cavities or the height of the vanes. To construct a comb SWS with hybrid dispersion characteristics, at least two comb SWSs with different slot depths should be employed, i.e. an SD SWS and a WD SWS. The SD SWS is characterized by a group velocity smaller than 0.1 times the speed of light, while the WD SWS is characterized by a group velocity $|v_g|$ larger than 0.1 times the speed of light. Figure 7.1 depicts an SWS with hybrid dispersion characteristics, showing (from left to right) the input rectangular waveguide, a three-period WD SWS, a three-period SD SWS, a three-period WD SWS, a three-period SD SWS, a three-period WD SWS, and the output rectangular waveguide. The SWS starts and ends with WD SWSs, as their weaker dispersion characteristics lead to a broader bandwidth. A cylindrical electron beam flows through the square beam tunnel drilled in the vanes. Here, we ignore the design procedures of the SD SWS and WD SWS and give the structural dimensions directly in table 7.1.

Once the dimensions of the structural elements are determined, the dispersion characteristics of the WD SWS and the SD SWS can be obtained through theoretical calculation or numerical simulation, as shown in figures 7.2(a) and (b). The criterion used to design the two SWSs is that the central synchronism frequencies should be the same at the operational beam voltage. In this example, both SWSs are designed to have the same frequency of 661.0 GHz, and the synchronism voltage is 15.9 kV. The difference lies in the phase shifts of the single-period SWSs. The phase shifts of the SD SWS and WD SWS are 375° and 391.2° at 661.0 GHz, corresponding to group velocities of 0.06c and 0.15c, respectively. The operational phase shift of the

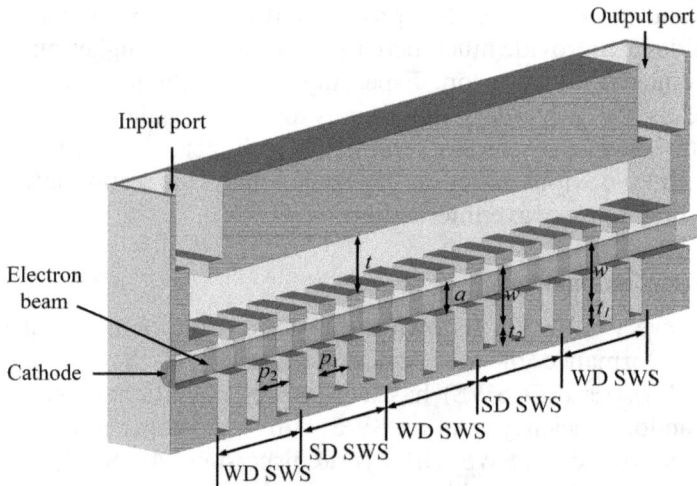

Figure 7.1. Cutaway view of the comb SWS consisting of three WD SWS sections and two SD SWS sections.

Table 7.1. The structural dimensions of the hybrid dispersion SWS.

Parameter	Value (mm)	Parameter	Value (mm)
p_1	0.115	t	0.15
l_1	0.09	w	0.15
p_2	0.12	d	0.05
l_2	0.11	h	0.28
a	0.09		

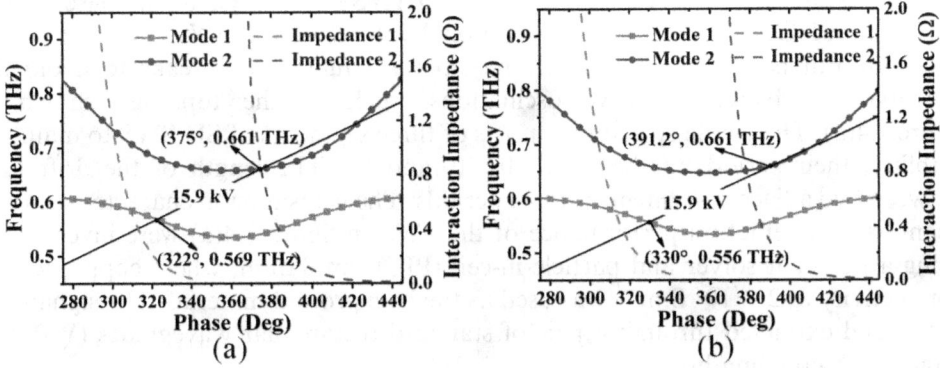

Figure 7.2. Dispersion and interaction impedance curves of (a) the SD SWS and (b) the WD SWS. © [2023] IEEE. Reprinted, with permission, from [6].

SD SWS is nearer to the cutoff frequency of the operational mode (mode 2); therefore, a higher beam–wave interaction impedance is obtained.

Figures 7.2(a) and (b) also show that there are intersection points between the beam voltage line and the fundamental modes at the backward wave regions, suggesting that there is a risk of backward wave oscillations for both SWSs. Fortunately, the beam–wave interaction impedances of the fundamental modes are much smaller than those of the operational modes. In addition, the corresponding frequencies of the intersection points are different, i.e. 569.1 GHz and 556.1 GHz; thus, if the numbers of sections of these two SWSs in figure 7.1 are less than the number of periods required to start oscillation, backward wave oscillations can be avoided.

The combination of the SWSs leads to changes in the dispersion characteristics of the overall SWS. Considering this SWS consists of a high dispersion section and a weak dispersion section with period numbers of k_1 and k_2, respectively, the equivalent dispersion of the HD SWS can be calculated. The average phase shift of the combined SWS is

$$\overline{\varphi} \approx \frac{[k_1\varphi_1 + k_2\varphi_2]}{k_1 + k_2}, \tag{7.1}$$

where φ_1 and φ_2 are the phase shifts of the single-period SD SWS and that of the WD SWS, respectively.

As φ_1 and φ_2 can be obtained from dispersion curves such as those shown in figure 7.2, the phase shift of the overall SWS at a given frequency can be calculated. The result is the equivalent dispersion curve, as shown in figure 7.3(b). The additional phase shift increases when the frequency approaches the cutoff frequency of the SD SWS (f_{1c}), as shown in figure 7.3(a).

If we select the period numbers for the SD SWS and WD SWS in each section to be 100 and 108, respectively, the equivalent dispersion curve of the overall SWS falls in between those of the SD SWS and the WD SWS, as shown in figure 7.3(b). That is to say, the overall SWS has an intermediate operational bandwidth compared to those of its sub-SWSs. As can be seen, the HD SWS has an equivalent phase shift of $383.4°$ at 661 GHz, indicating an operational voltage of 15.9 kV.

A uniform periodic SWS usually requires attenuators or breaks to avoid the potential for backward wave oscillations, such as the topology shown in figure 7.4(a). The whole structure consists of three segments of SD SWS to maintain stability; their period numbers are 70, 70, and 65. The length of the drift tube between two adjacent segments is three periods. The transmission characteristics and beam–wave interaction performance of the SWS in figure 7.4(a) were investigated using a transient solver and particle-in-cell (PIC) simulation, where copper with a conductivity of 2×10^7 S m^{-1} was used as the background material. The signals are fed in and extracted through a pair of standard rectangular waveguides (WR-1.5) attached to coupling holes.

In contrast, the HD SWS has the ability to suppress the oscillation in the absence of any break or attenuation, which makes it convenient to design an SWS in a single segment, as shown in figure 7.4(b). In this model, the five sections from the input port to the output port are: a 36-period WD SWS, a 50-period SD SWS, a 36-period WD SWS, a 50-period SD SWS, and a 36-period WD SWS, respectively.

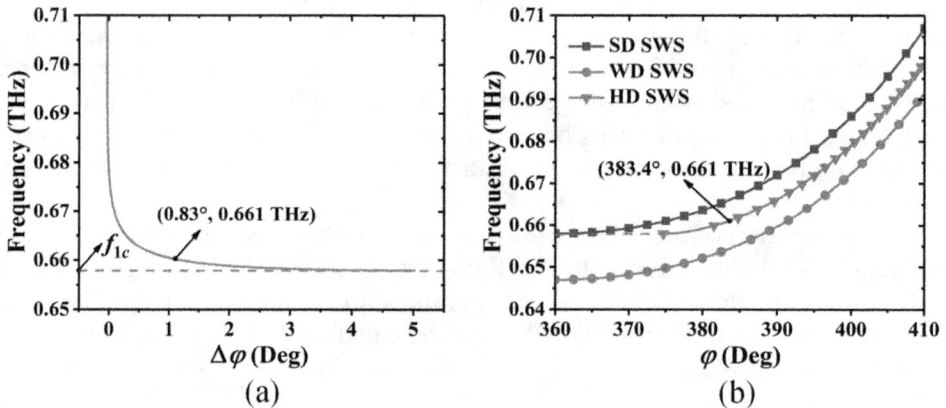

Figure 7.3. (a) Additional phase shift ($\Delta\varphi$) in the transitional region; (b) dispersion curves of the SD, WD, and HD SWSs. © [2023] IEEE. Reprinted, with permission, from [6].

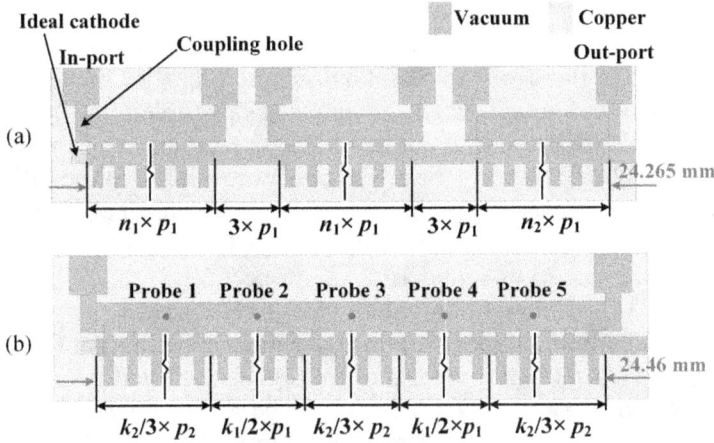

Figure 7.4. (a) Complete metal structures of the SD SWS with three segments and (b) the HD SWS with one segment. © [2023] IEEE. Reprinted, with permission, from [6].

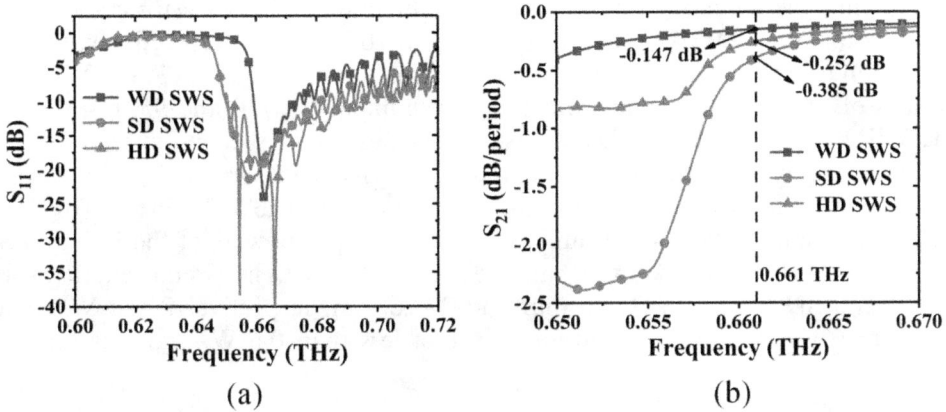

Figure 7.5. (a) S_{11} and (b) per-period S_{21} values of single-segment SD, WD, and HD SWSs. © [2023] IEEE. Reprinted, with permission, from [6].

The difference in dispersion characteristics leads to different transmission characteristics, including the cutoff frequency, reflection, insertion losses, etc. Figure 7.5 compares the reflection and insertion losses of different SWSs, where the period numbers used in the calculations for the single-segment SD SWS, the WD SWS, and the HD SWS are 70, 120, and 208, respectively.

In figure 7.5(a), the S_{11} parameter of the HD SWS is less than −15 dB in the frequency range of 657.1–668.8 GHz, and it is around −10 dB below the frequency of 660 GHz, which is near the cutoff frequency of the SD SWS. Therefore, in order to achieve a wide transmission bandwidth, a WD SWS is suggested for the input and output sections of the HD SWS. Taking the frequency point of 661 GHz as an example, the per-period S_{21} values for the WD SWS, HD SWS, and SD SWS are −0.147 dB, −0.252 dB, and −0.385 dB, respectively. While low loss is not always good, in the case where the period number is low in each section and the section number is large, low transmission losses lead to obvious reflections within those sections at the cutoff band of the SD SWS. Therefore, it is recommended to use fewer segments and as many cycles per segment as possible.

A comparison of the beam–wave interaction performances of the HD SWS, the SD SWS, and the WD SWS shows the advantage more clearly. In the PIC calculations, all the electric parameters were set to the same values each time, i.e. the input power was 1 mW, the focusing solenoid magnetic field was 0.6 T, and the beam radius, current, and voltage were 35 μm, 20 mA, and 16.0 kV, respectively.

Backward wave oscillations can usually be investigated using the PIC results, such as the oscillation starting period number, oscillating frequency, etc. Figures 7.6(a) and (b) show the spectra of the signals reflected from the input ports of the single-segment SD and WD SWSs. As can be seen, an obvious backward wave oscillation frequency appears when the period number exceeds 100 for the SD SWS or 130 for the WD SWS. Therefore, it is reasonable to suggest that the number of periods of the SD SWSs and WD SWSs within a segment should not exceed 90 and 120, respectively.

Applying the topology structures shown in figure 7.1 and setting the period numbers for the WD SWS and the SD SWS to 108 and 100, respectively, the PIC simulation results at 661 GHz can be obtained, as shown in figure 7.7. The WD SWS consists of two segments, which have period numbers of 110 and 95, respectively. The 3 dB bandwidth of the HD SWS is 4.7 GHz, as can be seen in figure 7.7(a). The improvement is around 51.6% compared to the SD SWS circuit. In addition, the maximum output power is 2.40 W, an improvement of 33.3%. Figure 7.7(b) shows the output–input curves of the three SWSs. The output powers of the HD and SD SWSs are almost saturated at an input power of 1 mW. The circuit lengths of the hybrid and SD SWSs are 24.46 mm and 24.26 mm, respectively. The WD SWS circuit produces the smallest output power of less than 0.6 W, and it is far from

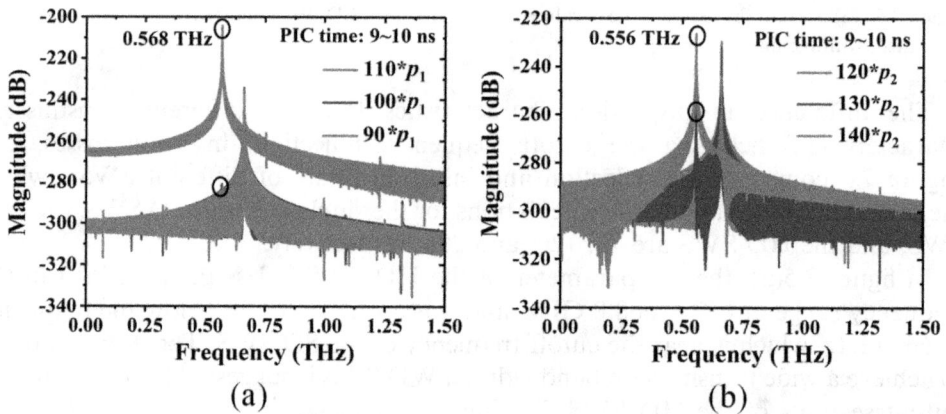

Figure 7.6. Spectrums of reflection signals at the input port of the SD SWS (a) and WD SWS circuits (b); both are single-segment circuits. © [2023] IEEE. Reprinted, with permission, from [6].

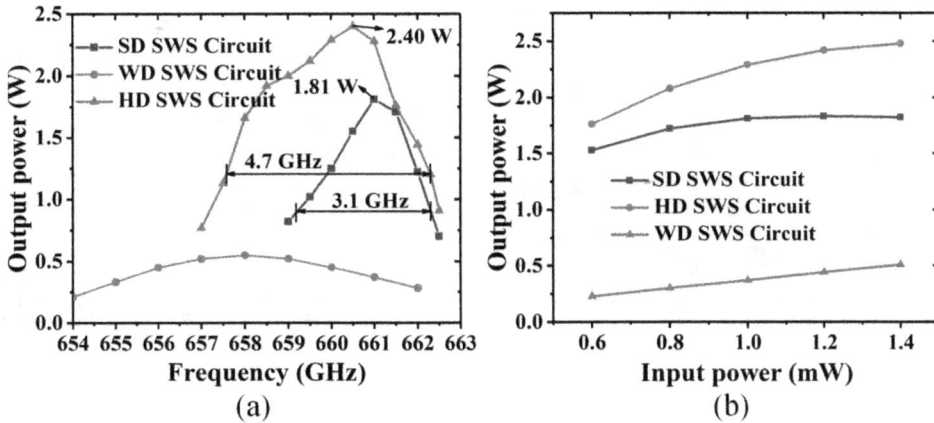

Figure 7.7. (a) Variation of output power with frequency; (b) variation of output power with input power at 661 GHz. © [2023] IEEE. Reprinted, with permission, from [6].

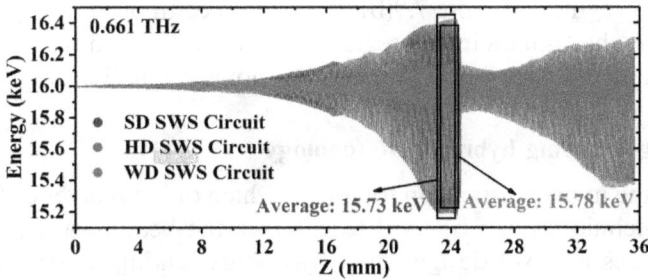

Figure 7.8. Phase-space diagrams of the three circuits at 661 GHz. © [2023] IEEE. Reprinted, with permission, from [6].

saturation, as seen in figure 7.7(b). Further research shows that it is saturated when the circuit employs three segments with period numbers of 100, 100, and 90, respectively, corresponding to an overall length of 35.52 mm, as shown in figure 7.8. However, as the beam tunnel is very small at the terahertz frequency band, an increase in length leads to problems in the implementation of the electron optical system.

However, further analysis of the power improvement of the HD SWS over the SD SWS shows there is no significant difference in the beam–wave transfer energy during the beam–wave interactions. Instead, the difference lies in the power extraction. The electrons in the SD SWS circuit lose more direct current energy to the high-frequency field compared to the HD SWS circuit, but the electrons then absorb more energy within the acceleration zones, as shown in figure 7.8. In other words, the SD SWS has a lower group velocity, i.e. a lower power flow, which is not conducive to the energy output.

The spectra of the output signals from both ports are shown in figure 7.9(a), which shows almost no other peak besides the input signal frequency of 661 GHz.

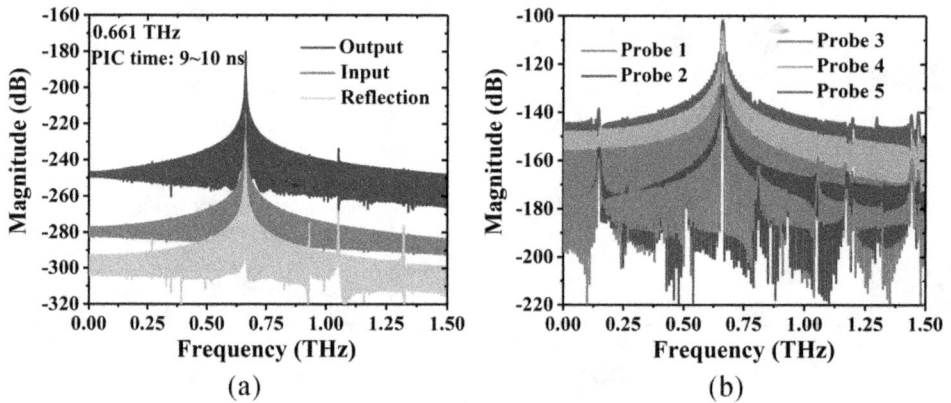

Figure 7.9. Spectra of (a) port signals; (b) readings from five electric field probes. © [2023] IEEE. Reprinted, with permission, from [6].

The probes from No. 1 to 5, as labeled in figure 7.4(b), are set to monitor the field increase in every segment. Figure 7.9(b) shows the spectra of the five probes; as can be seen, there are no harmonics in any segment. This is validation that the HD SWS has the ability to suppress backward wave and harmonic oscillations in a single segment.

7.1.2 Traveling–standing hybrid wave topology

In addition to the hybrid dispersion topology, which only involves different traveling waves, other techniques can be used to enhance the beam–wave interaction. This section introduces a novel design for a traveling–standing hybrid wave topology intended to enhance the gain and output power of terahertz traveling-wave tube (TWT) amplifiers [7]. The scheme consists of two SWS segments, which serve as broadband input and output segments, and at least one multigap resonant cavity, which serves as a beam-bunching segment. This combination results in a complementary hybrid of traveling-wave and standing-wave elements. It is expected to yield a traveling-wave tube amplifier with higher gain, higher power, and a more compact interaction length, even though its bandwidth may suffer compared to the bandwidth of a pure traveling-wave tube amplifier.

The verification of the traveling–standing wave hybrid scheme is based on a G-band TWT, as illustrated in figure 7.10. From input port to output port, the components used are a section of the classical staggered double-vane SWS including input/output ports, one or more ladder-type multigap resonant cavities without external ports, and another staggered double-vane SWS including input/output ports. These components share a common electron beam. The traveling-wave and standing-wave components are isolated from each other, since the electron beam tunnel is cut off for the operational electromagnetic wave signals. This is somewhat like a multicavity klystron, in which the electron beam gains high-frequency information due to modulation by the input signal; the premodulated electron beam is then further modulated by excited high-frequency fields in the resonant cavities; finally, the well-bunched electron beam passes through a second SWS and

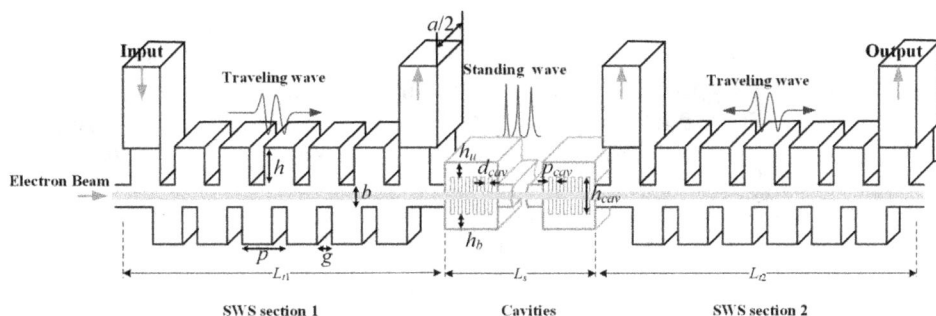

Figure 7.10. Schematic of the hybrid interaction circuit consisting of traveling-wave and standing-wave components. © [2020] IEEE. Reprinted, with permission, from [7].

excites a high-frequency electromagnetic field at the same frequency as the input signal, which is guided out through the broadband output structure.

However, while the similarity between the multisegment TWT and the multi-cavity klystron lies in their topology, there is a significant difference in the formation of their beam–wave interaction. In detail, there is a traveling wave on the SWS in a TWT: as the synchronization between the wave and beam is present through the whole SWS, beam modulation and bunching happen from the start to the end; in contrast, although the conventional klystron has standing waves, modulation happens only at the open end of the cavity.

If an WD SWS is selected, the operational bandwidth of a TWT can be quite wide, while if high-Q resonant cavities are selected, the energy transfer efficiency between the electron beam and the electromagnetic wave can be quite high while keeping the circuit short. In the traveling–standing wave topology, the idler cavities are supposed to serve as an enhancing element to increase the saturated energy transfer efficiency and reduce the overall length of the interaction circuit. The problem now is how to design the idler cavities to achieve synchronization between the traveling wave and the standing wave.

Although the principle behind the traveling–standing wave hybrid topology is simple and clear, an exquisite design is necessary to realize the expected performance improvement. In addition, the fewer the cavities, the more compact the structure and the better the design. The staggered double-vane SWS, which is well known for advantages such as a planar structure, suitability for a large current sheet electron beam, and the all-important wide bandwidth, is chosen as the input/output beam–traveling-wave interaction component. The ladder-type multigap resonant cavity, which is widely used in extended interaction klystrons (EIKs) due to its advantages such as high R/Q, is chosen as the beam–standing-wave interaction component. Here, R and Q denote the shunt resistance and the quality factor of the resonant cavity, respectively.

As the SWSs and the resonant cavities share a common electron beam, their designs must be coherent. An SWS can have a broad operational frequency band, while a resonant cavity can only provide an efficient response in a limited frequency band characterized by $\sim 1/Q$. Hence, defining the resonant frequency of the resonant cavity becomes the core problem.

In conventional klystrons, a popular way to improve the output cavity's performance is to set the idler cavity's impedance to be inductive at the operating frequency. That is to say, the cavity should have a higher resonant frequency (f_0) than the system's operating frequency (f) so that:

$$\frac{1}{Z_{\text{cav}}} = \frac{1}{R} + j\left(\frac{f}{f_0} - \frac{f_0}{f}\right)\frac{1}{R/Q} \tag{7.2}$$

where R/Q is the characteristic impedance defined as [8]:

$$\frac{R}{Q} = \frac{\left(\int_{-\infty}^{+\infty}|E_z|dz\right)^2}{2\omega W_s}. \tag{7.3}$$

Here, W_s and E_z are the stored energy and the electric field along the beam tunnel, respectively.

Starting from one idler cavity condition, it should be kept inductive over the required frequency band. The only way to do this is to set its resonant frequency (f_0) near the upper end of the required frequency band. For example, if the desired bandwidth is 10 GHz centered at 220 GHz, it is suggested to set $f_0 = 225$ GHz initially. The exact frequency will eventually be determined by the PIC simulation.

In practical scenarios, R/Q does not indicate the beam–wave interaction efficiency accurately enough. Another quantity known as the coupling coefficient M should also be investigated, which is defined as [8]

$$M = \frac{\left|\int_{-\infty}^{+\infty} E_z e^{j\beta_e z}dz\right|}{\int_{-\infty}^{+\infty} |E_z|dz} \tag{7.4}$$

where β_e is the propagation constant of the direct current electron beam.

For comparison, for a beam voltage of 21 kV, M is calculated to be 0.3494 for an axial electric field with an amplitude of 2.56×10^{10} V m^{-1}, and the effective interaction impedance $(R/Q) \times M^2$ is about 62.7 Ω. As can be seen, compared to the traveling-wave circuit, the standing-wave circuit has a much higher interaction impedance, even though the multigap resonant cavity has a shorter period ($p_{\text{cav}} = 0.37$ mm) than that of the SWS ($p = 0.44$ mm). In this case, the designed seven-gap cavity is 2.4 mm in length, which is equivalent to 5.5 periods of the SWS component. These lengths play important roles in the overall circuit length, which is why we discuss them.

As the operational frequency band of an SWS is much wider than that of a resonant cavity, it is impossible to balance the performance at both the upper- and lower-frequency ends using a single resonant cavity. Therefore, more idler cavities should be employed for a broader bandwidth. However, before this can be done, the SWS has to be redesigned to shift it toward the lower-frequency end.

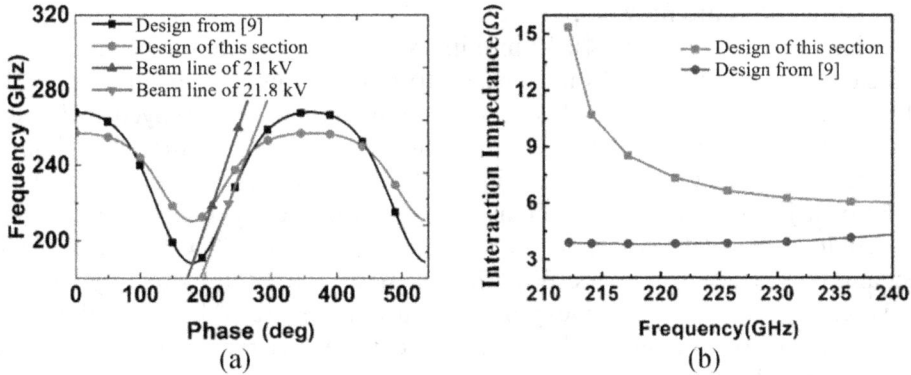

Figure 7.11. (a) Dispersion characteristics and (b) interaction impedance of the proposed staggered double-vane SWS and the reported wideband staggered double-vane SWS. © [2020] IEEE. Reprinted, with permission, from [7].

Table 7.2. The structural parameters of the circuit.

SWS		Cavity	
Parameter	Value (mm)	Parameter	Value (mm)
a	0.712	d	0.150
b	0.150	h_b	0.362
h	0.437	h_u	0.362
g	0.133	h_{cav}	0.30
p	0.440	p_{cav}	0.37

Figure 7.11 shows the dispersion characteristics and interaction impedance of the designed staggered double-vane SWS. To show the characteristic clearly, the dispersion characteristic curve of a wideband staggered double-vane SWS [9] is also presented. It can be seen that the staggered double-vane SWS has very flat dispersion characteristics, which is why it can, in principle, provide a bandwidth as broad as 50 GHz at around 220 GHz. An increase in the operating voltage shifts the operating point toward a lower cutoff frequency, leading to a higher interaction impedance but worse dispersion characteristics. We can now say that high-gain TWT operation can be achieved by increasing the interaction impedance.

Thus, the SWS is improved by operating it near its lower cutoff frequency, thereby obtaining a higher interaction impedance in the lower frequency band, while the worsening of its dispersion characteristics is acceptable because its bandwidth is still broad enough compared to the bandwidth of the resonant cavities. The optimized beam voltage is about 21 kV. In the case of the upper frequency band, the multigap cavities play a more important role due to their higher interaction impedance. Table 7.2 summarizes the typical dimensions of the designed staggered double-vane SWS.

PIC simulations are now employed to validate the concept. The sheet beam used in the simulations is 0.4 mm × 0.12 mm in cross-section, while the beam tunnel has dimensions of 0.45 mm × 0.15 mm. The electron beam voltage and current are 21 kV and 0.2 A, respectively. A 0.9 T solenoid magnetic field is employed to focus the sheet beam within the length of the overall high-frequency structure. The first SWS has 30 periods, while the second has 35 periods. Ohmic losses are included in the PIC simulation by enclosing the circuit in an oxygen-free copper (OFC) shell with a conductivity of $\sigma = 2.2 \times 10^7 \, \text{S m}^{-1}$, which is widely used at 220 GHz due to the skin effect and the surface roughness. With these settings, the resulting quality factors for the five-gap, seven-gap, and nine-gap resonant cavities are 503, 499, and 496.

The function of a single multigap cavity can be observed from the frequency response curve shown in figure 7.12, where the input power is set to 11.25 mW. It can be seen that the single multigap cavity shows a significant improvement in the output power when the SWS and resonant cavity are designed for a high degree of coherence. As shown in the figure, an increase in the number of gaps leads to an increase in output power. When a seven-gap cavity is employed, the maximum (unsaturated) gain can reach 40 dB at 219 GHz; however, the most important feature is that the overall length is only 32 mm. The results of the 3 dB gain–bandwidth product for resonant cavities with the different gap numbers shown in figure 7.12 are listed in table 7.3.

Figure 7.13 shows the frequency response curve of the hybrid SWS for an input power of 11.25 mW. It can be seen that the output power at the upper frequency end has been improved significantly, and an extended bandwidth is obtained. The saturation output power is 190 W at 220 GHz, corresponding to a gain of ~42.27 dB and an energy transfer efficiency of 4.5%. Driven by a solid-state power source providing an input power of 11.25 mW, the bandwidth at an output power exceeding 50 W can reach up to 8.5 GHz at around 220 GHz.

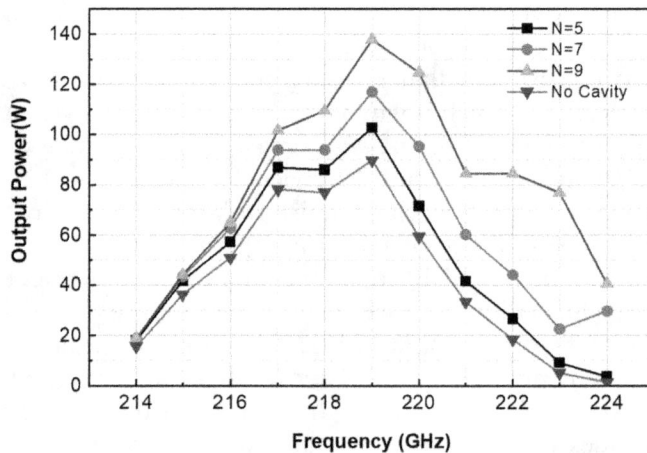

Figure 7.12. Output powers of the proposed TWT enhanced by cavities with different numbers of gaps. © [2020] IEEE. Reprinted, with permission, from [7].

Table 7.3. The (gain)*(3 dB bandwidth) product for each curve plotted in figure 7.12.

N	Gain (dB)	3 dB bandwidth (GHz)	Gain–bandwidth product (dB·GHz)
0	38.9	4	155
5	39.5	4.5	178
7	40.1	5	200
9	40.8	6.5	265

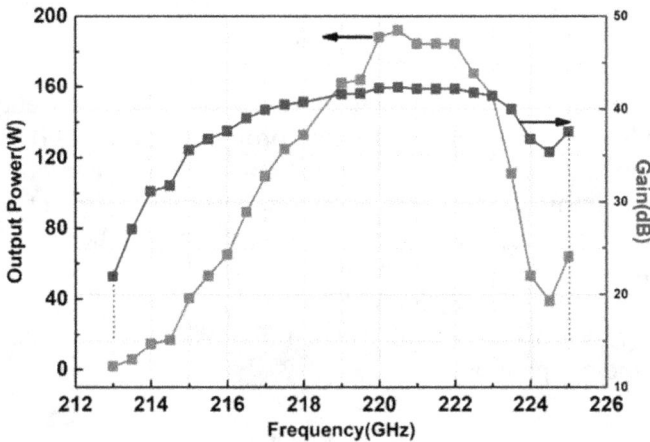

Figure 7.13. Output power and gain of the hybrid TWT. © [2020] IEEE. Reprinted, with permission, from [7].

As calculated from the port signals, 190 W is the total output power, including all harmonics. However, spectrum analysis proves that the fundamental harmonic contributes almost all of the output, which is about 187.5 W according to the FFT transform. Therefore, the fundamental harmonic is amplified with a gain and an electronic efficiency of 42.22 dB and 4.47%, respectively. There is now plenty of evidence to show that the novel traveling–standing wave hybrid topology is superior to conventional TWTs and EIKs.

7.2 Artificial-intelligence-aided design

In traditional design processes for a specified topological structure, the complexity of its structural and boundary conditions makes traditional electromagnetic simulation calculations, such as those performed by the CST Eigenmode Solver, time-consuming. Calculating the dispersion characteristics, coupling impedance, and cold test parameters corresponding to a specific set of structural parameters can take a considerable amount of time. Moreover, when working conditions change, designers have to repeat this intricate process.

Today, artificial intelligence (AI) has demonstrated powerful predictive capabilities in various fields. Deep neural networks (DNNs), in particular, are versatile network structures commonly used for learning and discovering complex patterns and features in data. They find wide applications in tasks such as image, audio, and text processing, as well as classification and regression. Therefore, leveraging the powerful learning capabilities and fast computational speed of DNNs is a promising avenue for predicting the dispersion characteristics and coupling impedances of structures [10–12]. While these optimization methods have accelerated the process of designing high-frequency structures, their optimization objectives only consider the dispersion characteristics of the fundamental mode, neglecting higher-order dispersion characteristics and coupling impedances. This implies that the optimization algorithms cannot assess the risk of higher-order standing-wave oscillations and backward-wave oscillations during beam–wave interaction. This necessitates a reconsideration of the forward mapping design to allow the algorithms to consider more performance parameters.

This section presents a rapid design process for a TWT that employs a folded waveguide SWS with gradient dispersion properties (FWG-GDP SWS). After specifying the topology, setting goals, and defining operating conditions, this process employs AI algorithms to automatically complete the design of an FWG-GDP SWS, enhancing the energy exchange between the electron beam and the traveling wave. This approach can significantly save the designer's time and quickly demonstrate the performance limits of the chosen topology. It is crucial to emphasize that this scheme is applicable to all SWSs that can be parameterized and exhibit periodic repetition.

7.2.1 Forward mapping algorithm

We now investigate the relationship between the high-frequency characteristics and the combination of the structural parameters. The single-period folded waveguide (FWG) SWS, as shown in figure 7.14, can be characterized by five parameters: the wide side of the waveguide a, the narrow side of the waveguide b, the height of the waveguide h, the half-period length of the waveguide p, and the radius of the electron beam channel r.

Training set preparation: the goal is to train the DNN to predict the dispersion characteristics and coupling impedance based on the input parameters a, b, h, and p for a given r. We collect the dispersion characteristics and electric field strength for different structural parameters and calculate the coupling impedance of the first six

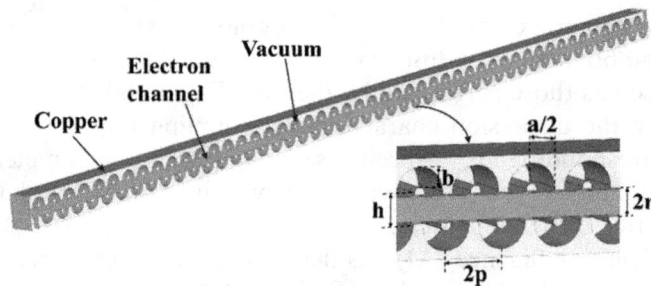

Figure 7.14. Diagram of the FWG SWS and a cylindrical electron beam.

eigenmodes using the eigenmode solver in CST Studio Suite. For each eigenmode, the phase differences of the periodic boundary range from 360° to 540° in steps of 5°, forming a mode frequency (MF) set with 6 × 37 points. The coupling impedance matrix of the first mode (Kc1) is represented by 1 × 37 points.

We select an SWS operating in the G band. The sample numbers for the different parameters are different, depending on their influence on the operating frequency. We now traverse all combinations, totaling 7040 samples, using an automated script and the CST Eigenmode Solver to complete data set sampling. We randomly split the data set into two halves: 50% is used for training, and the remaining 50% is reserved for validation to ensure that the trained network does not overfit. Since the output data number is $7 \times 37 = 259$, which is significantly greater than the four input parameters, we divide them into seven DNNs in order to reduce the model size and training difficulty. Each network has four inputs and 37 outputs, as illustrated in figure 7.15 and table 7.4.

For MF1 and Kc1, the model consists of an input layer with four input neurons, including three hidden layers and one output layer. Each hidden layer is connected to a rectified linear unit (ReLU) activation function to introduce the network's nonlinear capabilities. There are various choices of activation functions, including sigmoid, tanh, ReLU, PReLU, etc. The sigmoid and tanh functions are more suitable for classification problems and may face the issue of vanishing gradients in regression problems, especially in deep networks. Due to its simplicity, nonlinearity,

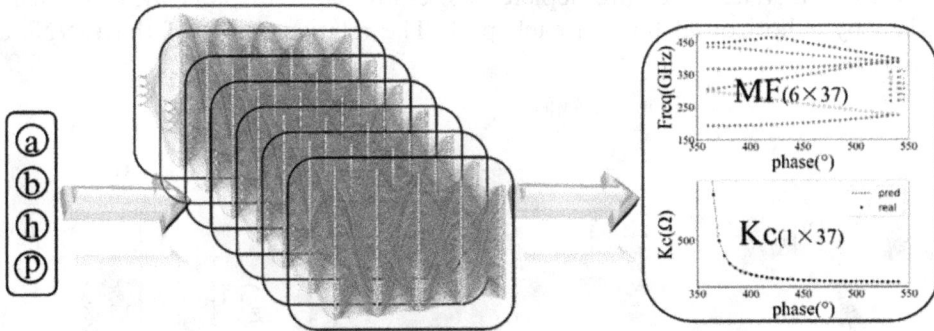

Figure 7.15. Schematic diagram of the overall network structure.

Table 7.4. The structural parameter ranges.

Structural parameter	Step (μm)	Range (μm)
a	50	(640,840)
b	10	(70,220)
h	22	(210,430)
p	20	$b+$ (60,200)

and sparse activation nature when only a subset of the neurons is activated, ReLU is one of the most popular activation functions. However, it has the issue of 'neuron death,' in which certain neurons may never be activated during training, leading to a failure to update their weights. Variants such as PReLU aim to mitigate the problem of neuron death but introduce additional complexity into the network training process. In the training process, balancing pros and cons, ReLU is sufficient for current use as the network's activation function.

In additional, data preprocessing is a crucial step before inputting data into a deep neural network. Preprocessing aims to ensure that the data has appropriate scales, distributions, and forms for effective learning by the network. Data preprocessing usually includes normalization, feature scaling, noise removal, handling missing data, etc. Normalization helps to maintain stable gradients during backpropagation and accelerates the model's training speed and predictive performance. In addition, due to CST run errors caused by external factors during the execution of automated scripts and anomalies in some sampled data, it is necessary to remove or fill in any problematic data points.

The absolute relative error of each network in the test set is shown in figure 7.16. The relative error expression is:

$$\text{Relative loss} = \frac{1}{n} \frac{|y - \hat{y}|}{y_{\text{mean}}} \times 100. \tag{7.5}$$

In figure 7.16(a), the violin plot vividly illustrates the distribution of absolute errors for different DNNs, which are depicted as semitransparent plots with the median marked by a horizontal line on each plot. The relative errors of the DNNs are

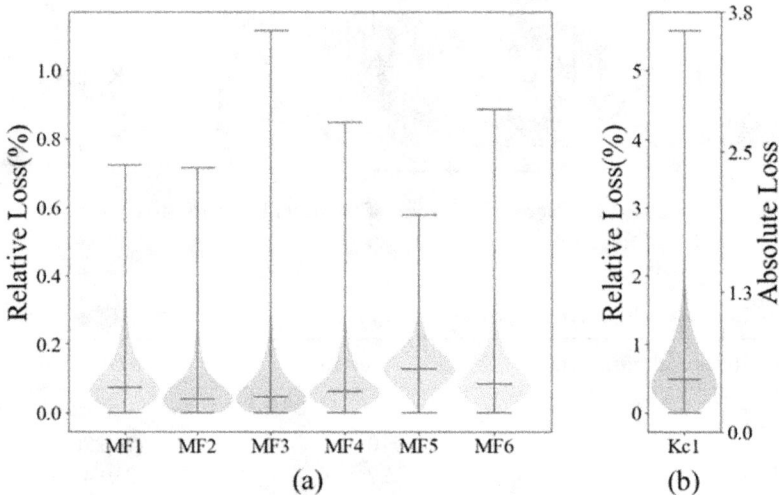

Figure 7.16. (a) Violin plots of the relative errors in the dispersion curves predicted by six DNNs and (b) the relative and absolute errors in the coupling impedance predicted by a DNN.

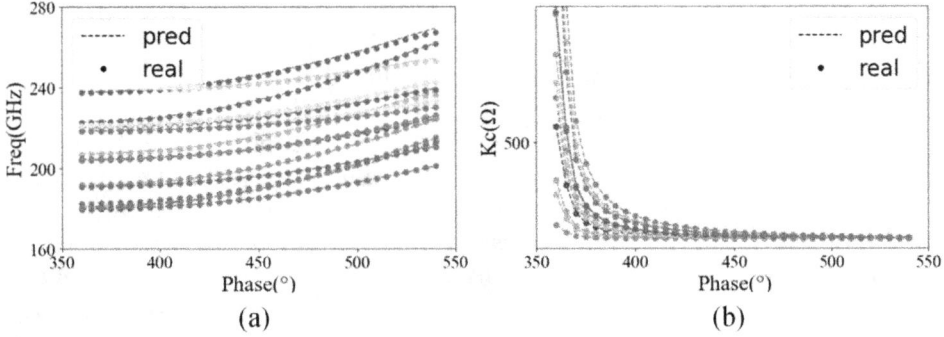

Figure 7.17. Comparison between DNN predictions and CST simulations for 15 random samples from the test set: (a) dispersion curves for mode 1, (b) coupling impedances for mode 1.

mostly less than 0.25% when predicting the dispersion characteristic for various structural parameters. Figure 7.16(b) represents the prediction results for the coupling impedance Kc1, which show relatively higher errors than the predicted dispersion characteristics. This is attributed to relatively large values occurring near the 2π point, as shown in figure 7.17. However, the absolute error remains within 3.8, showcasing the expressive capability of the DNN.

On a workstation with an i7-8700K processor (5 GHz) and an A4000 graphics card, the DNN can predict 3513 sets of test data within 6.8 s. In contrast, with the same computer hardware, the CST simulation software takes an average of 1961 s for one set of data. This signifies that the computational speed of the DNN is accelerated by a factor of 1.01×10^6.

7.2.2 Reverse search algorithm and uniform structure particle-in-cell verification

The cost function is defined as follows at the specified operating frequency:

$$\text{Cost} = \frac{0.02}{Kc_1} |U' - U| + \sum_{n=2}^{6} \frac{e^{-5v_{gn}}}{v_{gn}} \tag{7.6}$$

$$v_{gn} = \frac{\partial MF_n}{\partial \varphi} \bigg|_{MF_n = MF_{nc}} \tag{7.7}$$

where K_{c1} is the coupling impedance of the fundamental mode at the intersection point, U' is the predicted operating voltage (fundamental mode), U is the target operating voltage, and v_{gn} is the derivative of the dispersion with respect to the phase shift at the intersection point between the predicted beam voltage line and the nth high-order mode. When the beam voltage line intersects with the standing-wave point, the cost value tends toward infinity. When it intersects with the backward-wave region, the cost value is relatively large, and as the frequency increases (coupling impedance decreases), the cost gradually decreases. When the line intersects the traveling-wave region or there is no intersection, and the predicted voltage is close to the target voltage, the cost is close to zero.

Due to the four structural parameters of the curved waveguide and assuming the use of exhaustive enumeration, sampling each structural parameter in the range of ± 100 at 1 μm steps results in 1.6×10^9 possibilities. This implies the need to calculate the cold test parameters and determine the cost value for a vast range of structural parameters, making the process highly complex and inefficient. To overcome the issue of exhaustive enumeration, we opted for a search algorithm to efficiently find the optimal solution in this multidimensional parameter space. Commonly used optimization algorithms include the particle swarm optimization algorithm (PSOA), the genetic algorithm (GA), and simulated annealing (SA). In this work, we chose the PSOA due to its advantages such as simplicity, easy implementation, strong adaptability, and its ability to handle high-dimensional problems. In our implementation, we formalized the problem as an optimization problem for the PSO algorithm, in which each particle represented a combination of structural parameters. The fitness function was designed to calculate the cost value, and operations such as selection, crossover, and mutation simulated the movement of particles in the solution space. The PSO algorithm rapidly and effectively found the optimal structural parameters, improving the efficiency of optimization.

7.2.3 Rapid gradient dispersion property processing

Although the proposed algorithm can quickly obtain structural parameters that satisfy the waveguide interaction matching conditions and are less prone to standing-wave oscillations and backward-wave oscillations near the 2π and 4π points, increasing the number of periods in a uniform structure creates a risk of higher-order-mode backward-wave oscillations. This is because the working voltage line inevitably intersects the dispersion curve's backward-wave point near the $5\pi/2$ point. In cases with fewer periods, this point corresponds to a lower coupling impedance, reducing the likelihood of backward-wave oscillations.

In the following section, we outline a rapid processing method for designing an FWG-GDP SWS based on reverse search networks, as illustrated in figure 7.18. Among them, the mainstream particle beta is a beta representative that describes a position along the high-frequency structure. The particle distribution in the phase-space diagram is counted, and the two-dimensional histogram statistics are produced using the distance and the particle beta. The results are shown in figure 7.19. The yellow stars represent mainstream particle betas, which are the beta intervals with the largest number of particles in a certain position interval. The mainstream particle beta is the expected beta size for the design of the next SWS unit.

The working conditions are shown in table 7.5. We set $a = 750$ μm and $b = 135$ μm. Based on the variation process, we designed the FWG-GDP SWS as shown in figure 7.20. The black dotted line divides the whole high-frequency structure into several phase-velocity sections, which are consistent with the

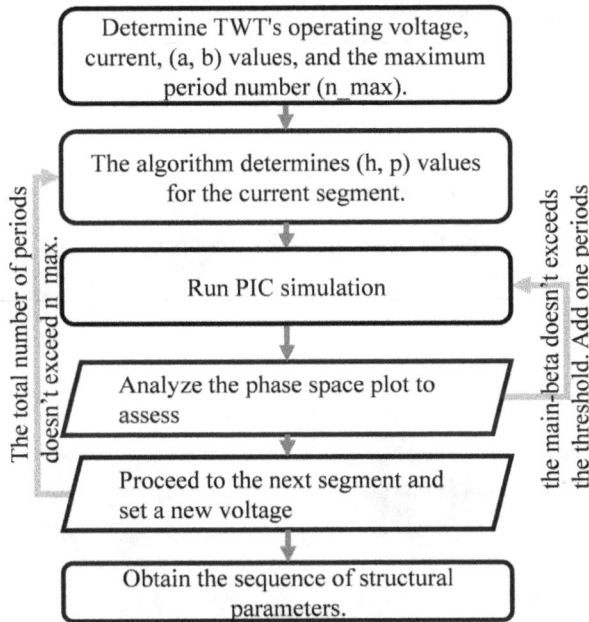

Figure 7.18. Rapid process for varying the phase velocity.

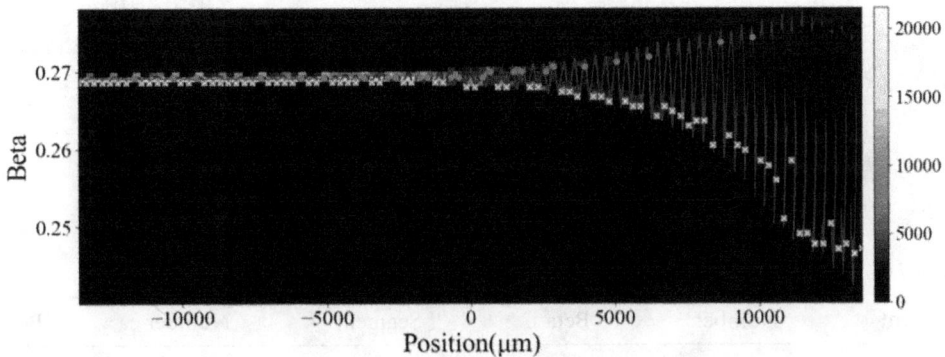

Figure 7.19. Two-dimensional statistical plot of electron beta.

upper tube. The red dotted line is the normalized phase-velocity beta of this segment. The beta and the number of periods for each segment are shown in table 7.6. The blue dots represent the phase-space diagram when the electron beam is stable.

When the input signal is 1 mW, the output signal and its spectrum obtained from PIC simulations are as shown in figures 7.21(a) and (b). As can be seen, the output average power is stable at 189.69 W, corresponding to a beam–wave interaction

Table 7.5. The initial working conditions.

Parameter	Value
Voltage	19.6 kV
Beam current	100 mA
Frequency	220 GHz
Focusing magnetic field	0.65 T
Conductivity of oxygen-free high thermal conductivity copper (OFHC)	5.8×10^7 s m^{-1}
Maximum number of periods	59

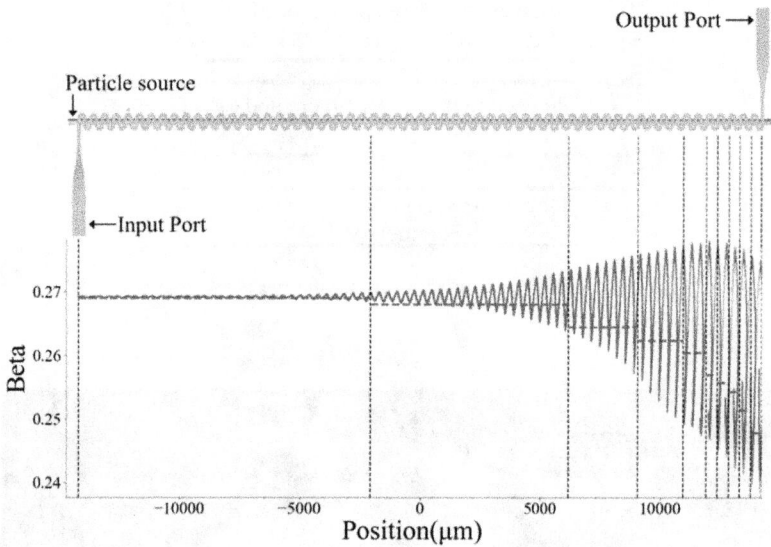

Figure 7.20. Two-dimensional statistical plot of the electron beta.

Table 7.6. The beta and number of periods per segment

Segment	Number	Beta	Segment	Number	Beta
1st	25	0.2690	6th	1	0.2569
2nd	17	0.2680	7th	1	0.2558
3rd	6	0.2644	8th	1	0.2543
4th	4	0.2623	9th	1	0.2514
5th	2	0.2605	10th	1	0.2478

efficiency of 9.67% and a gain of 52.8 dB. From the spectrum in figure 7.21(b), it is found that due to its high gain, there are higher harmonic components at 440 GHz and 660 GHz as well as the fundamental frequency at 220 GHz, but their amplitudes are extremely small; both are 40.32 dB lower than the amplitude at 220 GHz.

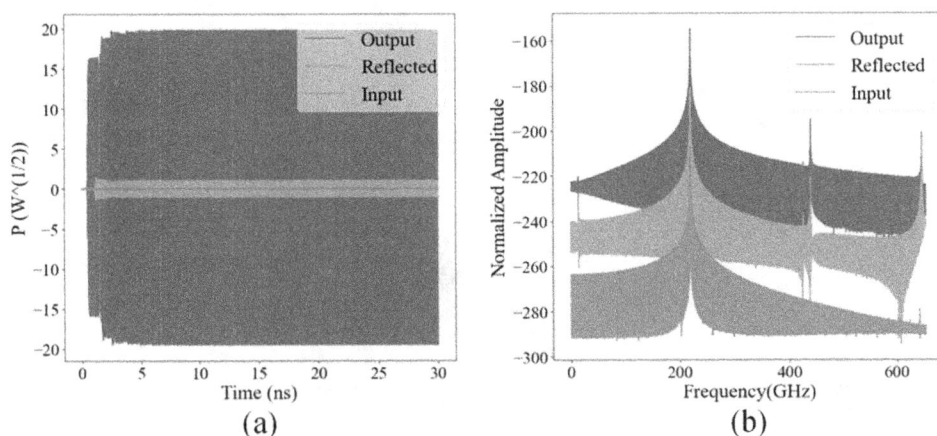

Figure 7.21. Signal and spectrum of its gradient structure. (a) Time domain; (b) frequency domain.

References

[1] Gamzina D, Himes L G, Barchfeld R, Zheng Y, Popovic B K, Paoloni C, Choi E and Luhmann N C 2016 Nano-CNC machining of sub-THz vacuum electron devices *IEEE Trans. Electron Dev.* **63** 4067–73

[2] Tucek J C, Basten M A, Gallagher D A and Kreischer K E 2016 Operation of a compact 1.03 THz power amplifier *Proc. IEEE Int. Vac. Electron. Conf. (IVEC) (Monterey, CA, USA)* pp 1–2

[3] Ives R L 2023 Advanced fabrication of vacuum electron devices *IEEE Trans. Electron Devices* **70** 2693–701

[4] Nguyen K T *et al* 2014 Design methodology and experimental verification of serpentine/folded-waveguide TWTs *IEEE Trans. Electron Devices* **61** 1679–86

[5] Booske J H 2008 Plasma physics and related challenges of millimeter-wave-to-terahertz and high-power microwave generation *Phys. Plasmas* **15** 055502-1–055502-16

[6] Dong Y *et al* 2023 A hybrid dispersion slow wave structure for 0.66 THz traveling wave tubes *IEEE Electron Device Lett.* **44** 1888–91

[7] Shi N J *et al* 2020 A novel scheme for gain and power enhancement of THz TWTs by extended interaction cavities *IEEE Trans. Electron Dev.* **67** 667–72

[8] Chodorow M and Wessel-Berg T 1961 A high-efficiency klystron with distributed interaction *IRE Trans. Electron Devices* **ED-8** 44–55

[9] Baig A *et al* 2017 Performance of a nano-CNC machined 220-GHz traveling wave tube amplifier *IEEE Trans. Electron Devices* **64** 2390–7

[10] Liu K, Xue Q, Zhao D and Feng J 2021 Intelligent forward-wave amplifier design with deep learning and genetic algorithm *IEEE Trans. Electron Devices* **68** 3568–75

[11] Zhu Y *et al* 2022 Inverse design of folded waveguide SWSs for application in TWTs Based on transfer learning of deep neural network *IEEE Trans. Plasma Sci.* **50** 3276–82

[12] Lan F *et al* 2023 Automated design of broadband folded-waveguide slow-wave structures for traveling-wave tubes via deep reinforcement learning *IEEE Trans. Electron Devices* **70** 3899–907

www.ingramcontent.com/pod-product-compliance
Lightning Source LLC
Chambersburg PA
CBHW071958220326
41599CB00032BA/6419